T0205591

FE Computation on Accuracy Fabrication of Ship and Offshore Structure Based on Processing Mechanics

Hong ZHOU · Jiangchao WANG

FE Computation on Accuracy Fabrication of Ship and Offshore Structure Based on Processing Mechanics

Science Press
Beijing

Springer

Hong ZHOU
School of Naval Arcitecture and Ocean
Engineering
Jiangsu University of Science
and Technology
Zhenjiang, Jiangsu, China

Jiangchao WANG
School of Naval Arcitecture and Ocean
Engineering
Huazhong University of Science
and Technology
Wuhan, Hubei, China

ISBN 978-981-16-4089-6   ISBN 978-981-16-4087-2   (eBook)
https://doi.org/10.1007/978-981-16-4087-2

Jointly published with Science Press, China
The print edition is not for sale in China Mainland. Customers from China Mainland please order the print book from: Science Press.
ISBN of the Co-Publisher's edition: 978-7-03-069518-5

This Springer imprint is published by the registered company Springer Nature Singapore Pte Ltd.
The registered company address is: 152 Beach Road, #21-01/04 Gateway East, Singapore 189721, Singapore

Preface

The construction of ship and offshore structures is a complicated process involving a series of thermal processing technologies, such as flame welding, plate bending with induction heating, and welding. The dimensional accuracy of ship and offshore structures, as well as their fabrication cost and schedule, can be significantly influenced by fabrication deformations caused by the elastic-plastic mechanical response during individual thermal processing. Two fabrication deformations typically occur, namely, in-plane shrinkage and out-of-plane bending deformation.

Owing to complex physical behaviors resulting from individual thermal processing, the engineering solution to these mechanical responses during the construction of ship and offshore structures is generally determined by experimental measurements and fabrication experience. With the fast increase of labor cost and requirement of production efficiency, it is of great significance to solve such mechanical problems holistically with scientific investigation through advanced computational approaches. This will enhance the technological level and construction quality of ship and offshore structures.

To address actual engineering problems during the construction of ship and offshore structures, advanced computational approaches such as thermal elastic-plastic and elastic finite element (FE) computations were developed and employed to examine physical behaviors. In addition, the generation mechanism of the mechanical response was also clarified. Thus, computational analysis was carried out not only for fabrication deformation prediction due to thermal processing, but also for reduction of deformation tolerance and guarantee of fabrication accuracy. Therefore, appropriate construction processing based on theoretical analysis and computational modeling is desired and provided for actual engineering applications to improve the construction quality.

Focused on processing mechanics, the target of this monograph was to examine various typical thermal processing approaches employed during the construction of ship and offshore structures. The book is organized as follows:

Chapter 1 introduces the background of fabrication processing during the construction of ship and offshore structures. It also summarizes the progress of fabrication processing mechanics for thermal procedures, such as flame cutting, plate bending with induction heating, and welding.

Chapter 2 systematically summarizes the fundamental theory and method of FE computation for processing mechanics investigation, including non-linear thermal elastic-plastic FE computation with advanced computational techniques, welding inherent deformation evaluation, elastic FE computation with inherent strain/deformation as mechanical loading, interface element for fitting procedure consideration, and welding induced buckling.

Chapter 3 introduces a heating source model that is applicable to the cutting and emulation of high-strength thick plates. Then, the thermal-physical characteristic parameters of the NV E690 leg plate are calculated by JMATPRO and revised according to the theory of metallographic. The optimal values of the parameters, i.e., heating source, cutting speed, and radius of the heating source model, that ensure that the gear plates are cut through with no edge collapses during cutting are calculated. The optimized process parameters can be used to study the distribution of residual stress in the gear plates during cutting, thereby providing theoretical guideline for improving the construction technology of offshore platform legs.

Chapter 4 presents high-frequency induction heating for plate bending with double curvature. Several experiments were conducted, and out-of-plane bending deformations were measured by three-coordinate measuring devices in advance. Thermal elastic-plastic FE computations were carried out to represent the mechanical response and predict out-of-plane bending deformation. Then, the bending moment was evaluated from elastic FE computation. Good agreement was eventually observed between the computed results and the corresponding measurements.

Chapter 5 focuses on the out-of-plane welding distortion of fillet welded joints and stiffened welded structures with experimental measurements and FE computations. Fillet welded joint considered as typical welding during the construction of ship and offshore structures was practiced to measure the welding distortion, while thermal elastic-plastic FE analysis was carried out for computational accuracy validation. Moreover, stiffened welded structure deemed as the typical ship panel with fillet welding was experimentally and numerically investigated. To improve the computational efficiency, elastic FE computation was carried out with welding inherent deformation as mechanical loading. The computed results were in good agreement with measurement data, although with loss of computational accuracy.

Chapter 6 examines the out-of-plane welding distortion during the fabrication of ship structures. Hatch coaming structure fabrication of bulk-cargo ship was first examined. Then, out-of-plane welding distortion of fillet welding was measured and predicted with thermal elastic-plastic FE computation. Finally, welding inherent deformation was evaluated from elastic FE analysis. The fabrication accuracy could be enhanced by considering an improved manufacturing processing. Lightweight fabrication of ship panel was then examined with combined thermal elastic-plastic FE computation for typical welded joints and elastic FE computation for the actual ship panels. Welding induced buckling was investigated through elastic FE computation by employing the large deformation theory. Welding inherent in-plane shrinkage was the dominant cause that determined the buckling modes. Bending deformation was considered a disturbance that triggered buckling behavior when the critical buckling condition was reached. Mitigation practice with intermittent welding procedure was theoretically and numerically conducted and examined for welding-induced buckling prevention. Major welded structures of container ships such as watertight transverse bulkhead and torsional box structures were then examined for accuracy fabrication. Out-of-plane welding distortion was measured by total statin during actual fabrication. The computed results of elastic FE analysis based on a welding inherent deformation database and the measurements of welding distortion showed good agreement. For the reduction of out-of-plane welding distortion, the welding sequence influence was computationally considered for accurate fabrication of the watertight transverse bulkhead structure. Optimization of the welded joint design was also examined for accurate fabrication of the torsional box structure.

Chapter 7 investigates the fabrication accuracy of offshore structures. Fast thermal elastic-plastic FE analysis with ISM and parallel computation was carried out to examine the mechanical response and welding distortion during fabrication of a cylindrical leg structure. Bead-on-plate welding was conducted to offset the dimensional tolerance due to the fabrication welding process. Elastic FE analysis with welding inherent deformation was also applied to predict the welding distortion of a cantilever beam structure, while practical techniques with opposite deformation, welding sequence modification, and fixture constrain were numerically examined to ensure fabrication accuracy.

The authors acknowledge the financial support of national natural scientific funding and collaboration projects, the experimental and measurement data from shipyard, and other scholars for research discussion and writing correction.

Zhenjiang, China   Hong ZHOU
Wuhan, China   Jiangchao WANG

Contents

Chapter 1
Introduction

The construction of naval architectures and marine structures is a complicated fabrication process involving cutting, plate bending, and assembling through thermal procedures. To enhance the fabrication accuracy and market competition, it is desired to optimize the processing parameters during thermal fabrication of shipbuilding through investigation of processing mechanics and clarification of mechanical response.

Based on mechanical theory, processing mechanics with thermal procedures focuses on the mechanical response during fabrication to improve the construction level of ship structures and optimize processing parameters. Thermal fabrication procedures such as cutting with oxygen, plate bending with natural gas, and arc welding constitute critical research issues given that they play important roles during the ship construction and all of them undergo non-linear thermal-elastic–plastic response. To address the thermal-elastic–plastic problem during the application of thermal fabrication procedures, finite element (FE) analysis is usually employed to obtain numerical solutions by considering processing conditions as thermal loading, as well as thermal and mechanical boundary conditions during computation. Therefore, it is useful and helpful to understand the generation mechanism of residual stress and distortion during the application of thermal fabrication procedures through mathematical modeling and FE computation. Currently, the processing parameters can also be assessed before actual ship construction. Moreover, an optimized fabrication plan can be considered to avoid the engineering problems of stress concentration and dimensional accuracy and to enhance the manufacturing quality significantly.

1.1 Research Background

With the rapid development of the world economy, there is an increasing demand for energy. The development of the seabed oil field all over the world has expanded

H. ZHOU and J. WANG, *FE Computation on Accuracy Fabrication of Ship and Offshore Structure Based on Processing Mechanics*, https://doi.org/10.1007/978-981-16-4087-2_1

from shallow sea to deep sea and even ice-sea areas. Consequently, the material and technical requirements for manufacturing of offshore platforms steadily grow, especially those of offshore drilling jack-up platform legs, which have already been constructed by Z-direction steel with yield strength over 690 MPa and maximum thickness of 210 mm [1]. Cutting a high-strength large-thickness rack is a complex thermal processing procedure involving heat transfer, material metallurgy, solid and fluid mechanics, and several other sciences, in addition to the complex interaction between materials and cutting gas. After cutting, the shrinkage stress is produced with the natural cooling down of the steel plate. Notable complex deformation may also be produced. As the first processing procedure of welding production, cutting efficiency and quality directly influence the welding deformation of the entire structure. The neglect of the residual stress and strain in turn influence the resulting welding quality and reduce the structure strength, with an underlying hidden danger of crack expansion. Flame cutting is a common method for plate cutting. However, dry carbides are energy-consuming, and the storage and use of acetylene are potentially hazardous. Given that the cutting process has numerous thermal-impact factors such as residual stress and deformation, cutting a large high-strength thickness plate for an offshore jack-up platform leg rack should be previously examined. Oxygen cutting is more complex. The cutting process will produce a non-uniform temperature field and will be accompanied by thermal strain and partial plastic deformation. Currently, there is no reference related to this area. Thus, thermodynamic behavior research on cutting large high-strength thickness racks demand a great deal of experiments to provide theoretical and technical support for the improvement of the construction process, which plays an important theoretical research role and has engineering application value.

After applying a cutting procedure, plate bending is usually employed for curve-plate processing during the construction of modern ship and offshore structures. The ship hull is made up of a large number of plates with complex curvatures, especially at the bow and stern, where many double-curvature plates such as saddle or sail shapes are located. Thus, plate formation is an essential procedure for shipbuilding, which will be related to the production efficiency and precision of the ship structure, as well as to the fabrication cost and schedule of shipyards [2]. Hot formation is an important method of plate bending in shipyards. The temperature distribution in the thickness direction is uneven because of local plate heating. Hence, a bending moment would be generated for the plate to produce out-of-plane displacement. There are three main procedures of hot formation according to the applied heating source, namely line heating with flame (oxygen-acetylene flame), laser heating, and high-frequency induction heating. At present, the first two procedures are more common than plate formation by high-frequency induction heating, which is still on an early application stage in shipyards. However, owing to its unique advantages, it has progressively attracted more attention. In contrast with plate formation by line heating with oxygen-acetylene flame, high-frequency induction heating further facilitates an accurate control of the heating region and automation. It also exhibits high efficiency and produces less pollution [3]. Moreover, in contrast with plate formation by laser heating, high-frequency induction heating has low cost, simple equipment, and large heating power. Overall, plate formation by high-frequency induction heating has a broad potential for efficient and automatic plate bending formation in shipyards.

To assemble different parts and components in a ship section, welding is generally and widely applied because it is a highly productive and practical joining method in construction and manufacturing industries, e.g., shipbuilding. However, it is also well-known that there are some difficulties with welding and welded structures, such as residual stress and welding distortion [4]. Welding distortion is one of the most complex difficulties in the fabrication of welded structures. It results from non-uniform expansion/contraction and the associated plastic strain produced in the weld and surrounding base material, which is in turn caused by the heating and cooling cycles in the welding process [5]. Basically, there are two major causes of welding distortion in welded structures: the inherent strain/deformation due to expansion and contraction, and the gap and misalignment that are produced in the joint before and/or during fitting and welding. Welding distortion degrades the performance of a welded structure owing to the loss of structural integrity and dimensional accuracy. Welding distortion mitigation delays the production schedule and increases the fabrication cost [6]. At present, it is impossible to eliminate or correct welding distortion completely because of the non-linear irreversible nature of the welding and straightening processes. However, welding distortion should and can be minimized to an acceptable magnitude to control dimensional tolerances during assembly. To mitigate welding distortion, it is necessary to predict the distortion of a weldment before a single bead is laid. Prediction of welding distortion should be considered part of the design cycle rather than part of the manufacturing cycle [7]. Therefore, prediction and mitigation of welding distortion are of critical importance in modern construction and manufacturing industries.

In advanced fabrication processes, ensuring dimensional accuracy by minimizing the possible welding distortion is required not only in the production stage but also in the design stage. If welding distortion under a given design and production processes can be predicted with enough accuracy, the proposed configuration for assembling a welded structure can be assessed and welding distortion mitigation can be achieved more efficiently.

1.2 Literature and Research Progress

Concerning the fabrication of ship and offshore structures, processing mechanics during flame cutting together with plate bending by induction heating and welding, were already examined with experimental and computational approaches. Related research studies focused not only on abnormal ship steel and stiffened panel structures but also on high-tensile strength steel and complex structures. We review these studies next.

1.2.1 Steel Cutting with Flame Heating

Steel cutting during shipbuilding was widely studied with both experimental and numerical approaches. Both domestic and international studies mainly addressed numerical simulations of laser cutting. Fewer studies conducted numerical calculations and simulations of oxygen cutting, mostly limited to research on the temperature field, stress distribution conditions, and process measurements [8–16]. Reddy [17], Dabby and Paek [18], Bunting and Cornfield [19], Modest and Abakian [20], and Basu and Srinivasan [17] carried out studies focused on the effects of heat characteristics and material properties on the quality of cutting processes. Modest and Abakians [20, 21] presented a mathematical model describing the process of material removal from the surface subjected to a high-intensity laser beam. O'Neill and Gabzdyl [22] presented a study on a laser-assisted oxygen cutting process that was capable of cutting sections up to 50-mm thick with power levels below 2 kW while maintaining excellent cut quality. The preliminary results were examined using both theoretical and experimental analyses. Gross et al. [23–25] also developed a mathematical model of laser cutting. They pointed out that energy transfer during a steel melting process exhibits a complex behavior that includes advection, evaporation, and radiation. A source model for laser beam and heat convective heat was also reported.

A three-dimensional heat-transfer model based on a finite difference method was developed by Mazumder and Steen [26]. This model allowed for the formation of a keyhole by considering the grid points within the keyhole as part of the conducting network but operating at high temperatures after evaporation. Kim et al. [27] formulated a two-dimensional model with triangular finite elements for the transient analysis of material removal by evaporation with a high-power laser. Yu [28] presented numerical studies of laser cutting by using three-dimensional finite element modeling. The phenomena of changing boundary and loading conditions, and phase changes during the cutting process were incorporated in the model. Kim [29] developed a three-dimensional computational model to analyze an evaporative laser-cutting process based on a steady-heat-conduction equation with constant laser velocity. Numerical results about groove shapes and temperature distributions were presented and compared with those derived from semi-analytical methods.

Most of the reported analytical studies are based on the solution of the heat transfer equation, which considers the effects of heat source movement. Numerical cutting models can be classified into two categories: those that predict the processing parameters from a set of input conditions, and those that model the behavior of the process to provide a better understanding.

The oxygen-cutting process is widely used in the industry as a thermal cutting process that can cut through steel plates of thicknesses ranging from 0.5 to 2,500 mm. The required equipment is low-cost and can be used manually or mechanically. However, in contrast with other steel cutting methods, most of the energy produced during the oxygen-cutting process comes from metal combustion rather than from metal melting or metal evaporation (see Powell et al. [30]).

Most studies on cutting temperature field adopted the conclusions published by various welding studies. The oxygen-cutting process is much more complex than welding or other cutting processes. First, there are two heat sources and slag removal implies a substantial amount of energy. Because of the cutting slot, the continuity of mass and heat is disrupted, a phenomenon that is not easy to model. To date, there are only a few analytical or numerical studies on composite heat source.

1.2.2 Plate Bending with High-Frequency Induction Heating

Induction heating techniques are usually employed in advanced manufacturing for plate bending, such as in shipbuilding. Many preliminary studies on plate formation with induction heating were conducted. Jang et al. [31] carried out a series of induction heating experiments under various heating conditions established in terms of plate thickness, heating speed and power, etc. The transient thermal distribution and final configuration of the examined plates were measured. Shen et al. [32] developed a mathematical model to examine the temperature field during high-frequency induction heating. The effects of heating parameters such as the distance between the plate and the coil, applied current, frequency, and turns of the coil on temperature profiles were also discussed. Zhang et al. [33] employed finite element analysis for hull plate formation with high frequency induction heating. The computed results were experimentally verified. Lee et al. [34, 35] carried out an experimental line heating with high-frequency induction heating equipment to examine the permanent deformation behavior of an SS400 thick plate. They observed that permanent vertical deformation increased with an increase of input power regardless of the deformed shape with saddle-type dual curvature.

Employing a triangle heating technique, Bae et al. [36] examined the steel-plate formation process with high-frequency induction heating. The computed deformations of steel plate were in good agreement with the experiments. Lee and Hwang [37] employed an automatic high-frequency induction heating with a triangular heating process for thick steel-plate bending. The influence of designing heating patterns on the out-of-plane bending deformation of an SS400 thick plate was examined. According to an experimental test, Park et al. [38] pointed out that the angular distortion behavior of plates undergoing high-frequency induction heating can be defined as a function of heat intensity and rigidity of the heated plate, in particular the size of the test specimen. For industrial application, Park et al. [39] presented an automated thermal formation system with induction heating for shipbuilding. It comprised scanning devices for measurement of plates, a high-frequency induction heater for plate formation, a computational unit for heating information, and a gantry robot with six axes. Zhang et al. [40] investigated the non-dimensional relation between technological parameters and thermal formation behavior of the ship hull plate with induction heating. An orthogonal experimental test was conducted to validate the critical value of technological parameters, such as breadth and depth of the heat affected zone (HAZ). To improve the computational efficiency, elastic FE analysis with bending moment was proposed to consider the flame heating with different line heating patterns [41].

Bae et al. [42] computed the heating flux of plate during induction heating. This simplified the coupling process between the electromagnetic field and temperature field. Zhang [43] proved that a plate with induction heating can result in the same transverse shrinkage and deformation as in flame heating through a series of experiments and calculations. The residual stress of the plate formed with induction heating becomes evenly distributed in the heating zone. Lee [44] proposed an efficient method for predicting plate deformation by high-frequency induction heating through multi-divisional analysis. Zhang et al. [45] investigated the effect of plate edge shrinkage on steel-plate formation. Their results showed that heating such that the inductor moves out of the plate edge is an effective technology that meets the formation requirement of a sail-type plate. Dong et al. [46] researched the effect of coil width on deformed shape and processing efficiency during ship hull formation by induction heating. As the heating width expands with the expansion of the induction coil width, the number of processing lines via line heating is reduced, thereby improving the processing efficiency.

1.2.3   Welding Distortion Prediction in Shipbuilding

After welding, welding distortion is inevitably generated. Flame heating such as spot and line heating is usually employed as a mitigation approach for straightening owing to its flexible operation. Moshaiov and Song [47] discussed residual distortion of the regions near and far away from the heated area with welding and flame heating by means of an approximate method. The near field was subjected to thermal elastic–plastic mechanics whereas the far-away field behaved elastically. The role of thermal loading in the base and the weld metal by welding and flame heating was discussed. Park et al. [48] examined the amending welding distortion of thin-plate structures with spot heating and line heating methods. The evaluations of temperature distribution of spot and line heating were carried out using FE analysis and practical experiments for various heating time intervals. Radial shrinkage and angular distortion caused by spot heating were also evaluated and experimentally validated. Wang et al. [49] focused on the buckling deformation of thin-plate welded structures with a computational analysis comprising two steps: (1) a three-dimensional thermal elastic–plastic FE analysis for inherent strain/deformation evaluation and (2) an elastic FE analysis employing inherent deformation previously evaluated. After predicting out-of-plane welding distortion, the same elastic FE analysis was carried out again to investigate the line heating for controlling and reducing the welding-induced buckling deformation. Later, Wang et al. [50] examined the out-of-plane welding distortion in ship panel structure production through prediction and mitigation with elastic FE computation. The computed results showed that the considered ship panel buckled near the edge and only bended in the internal region after welding. Then, straightening with line heating was numerically investigated. In the internal region, fast line heating was applied on the opposite side of welded joints to generate inverse bending for correcting the welding bending deformation. By contrast, slow line heating was applied to the edge regions to balance the welding-induced inherent strain for welding buckling mitigation. Blandon et al. [51] pointed out that although line heating is effective on reducing welding distortion, the operation still relies on empirical knowledge. They focused on reducing the welding angular distortion of a stiffened panel with three different gas heating patterns, namely a single heating line, two heating lines, and zigzag heating lines. A heat source model to represent the heat flux given by a gas heating torch was proposed and employed for evaluation of temperature distribution and out-of-plane welding distortion to select the most efficient gas heating pattern from the considered cases.

1.2.3.1   Advanced FE Computation for Welded Structures

Brown and Song [52] employed a solid-shell finite element model of a circular cylinder and a ring stiffened structure to investigate the effect of welding configurations, such as weld gap, clearance, and fixture, on the dimensional precision of the considered large structure. Later, Michaleris and Debiccari [53] presented a numerical analysis combining 2D welding computation with 3D structural analyses that is effective when the computer resources are limited. Jung and Tsai [54] examined the relationship between cumulative plastic strains and welding distortion in advance, and then mapped each cumulative plastic strain component evaluated by thermal elastic–plastic analysis incorporating the effects of moving heat and nonlinear material properties into elastic FE analysis using equivalent thermal strains. Jung [55] also predicted the welding distortion in the fabrication of ship panels by means of elastic FE analysis a with shell-element model in which only longitudinal and transverse plastic strains were employed to compute the welding distortion. Good agreement was observed with results of thermal elastic–plastic FE analysis. Zhang et al. [56] evaluated six components of plastic strain of each welding pass with thermal elastic–plastic FE analysis while moving the heat source. A solid-element model with shorter length was employed. Then, these plastic strain components were mapped to a shell-element model of a real welded structure for welding distortion prediction. Yang et al. [57] developed a stepwise computational approach to predict welding distortion of a large welded structure for enhanced computational efficiency. Lump-pass modeling of welded joints with thermal–mechanical analysis was carried out to obtain the plastic strain. Then, welding distortion prediction was achieved by mapping the plastic strains previously evaluated into the shell-element model of an actual welded structure.

For complex welded structures, stiffeners are popular to enhance the rigidity of whole structures, which are usually assembled with fillet welding. Michaleris and DeBiccari [58] predicted the welding distortion and welding-induced buckling of a stiffened ship panel mock-up with a numerical analysis technique that combined two-dimensional welding simulations with three-dimensional structural analyses in a decoupled approach. These numerical predictions can be utilized as a design tool to consider the effect of the welding procedures on the evaluation and optimization of the design configurations. For a fixed design, the predictions can also be utilized as a manufacturing analysis tool to evaluate the welding distortion with different welding processes and procedures for optimization analysis. Tsai et al. [59] studied the warping of welding thin-plate stiffened structures, in particular, welding distortion behaviors, including local plate bending and buckling as well as global girder bending. They pointed out that warping was primarily caused by angular bending of the plate itself and that buckling did not occur in the examined structures for a skin plate thickness greater than 1.6 mm unless the stiffening girder bended excessively. A FE method was employed to examine the distortion mechanism and the effect of welding sequence on panel distortion based on the joint rigidity method. Vanli and Michaleris [60] presented a welding distortion analysis approach for T stiffeners with a particular emphasis on welding-induced buckling instabilities. Two-dimensional thermomechanical welding-process simulations were performed to determine the residual stress and angular distortion. Then, the critical buckling condition together with the corresponding buckling mode and bowing distortion were computed in a three-dimensional eigenvalue and linear stress analysis. The impact of some factors such as stiffener geometry, weld sequence, weld heat input, and mechanical fixture on the occurrence of buckling and the welding distortion pattern were also investigated. Camilleri et al. [61] focused on the prediction and control of thermal distortion in the fabrication of multiply stiffened welded structures. A comprehensive simulation tool was developed and validated by welding distortion prediction, in particular the out-of-plane deformation generated in double-side fillet welded attachments. This simulation tool was employed to optimize the relative positions of a twin-arc configuration to obtain the minimum out-of-plane welding distortion for a single stiffener and double-fillet attachments, in which the close spacing of the welding arcs could trigger drastic out-of-plane welding distortion due to the welding buckling. Later, Camilleri et al. [62] developed and validated a wide-range simulation tool to predict welding distortion in stiffened plates and shells, with particular emphasis on out-of-plane welding distortion. A comparison of temperature profile and out-of-plane welding distortion with measurements confirmed the efficiency of the proposed algorithms for linking the thermal welding strains to the elastic–plastic structural response of the welded assembly by a static single-load-step analysis.

Jang et al. [63] applied an efficient computational approach to predict the welding deformation of stiffened panels with FE analysis employing inherent strain, which is defined as the residual plastic strain after the weld heat cycle. For the accurate evaluation of inherent strain in real welded structures, the degree of restraint change according to the different fabrication stages should be considered. Huang et al. [64] investigated the welding distortion of stiffened panel structures of naval ships, which were built with relatively thin plates and required a uniform surface to maximize the hydrodynamic performance and minimize the radar signature. Dimensional variation through the entire fabrication process in a production environment, in particular welding distortion, was assessed with measurements and advanced computational tools. The underlying distortion mechanism and critical process parameters in the examined stiffened panel structures were reported together with some other major findings. Deng et al. [65] pointed out that the prediction and reduction of welding distortion are critical to improve the quality of welded structures. They developed an elastic FE method to predict the welding distortion of large stiffened welded structures accurately with both longitudinal and transverse stiffeners during an assembling process considering both local shrinkage and root gap. The efficiency of the proposed elastic FE analysis was experimentally confirmed, and the influence of the initial root gap on the magnitude of welding distortion was investigated. Biswas and Mandal [66] examined the welding distortion patterns of large orthogonally stiffened plate panels with a numerical analysis methodology based on the quasi-stationary nature of welding. The effect of filler metal deposition was taken into account by implementing the so-called element birth-and-death technique. Biswas and Mandal [67] also presented two different modeling approaches for welding distortion prediction when assembling stiffened plate panels to address the high time consumption of conventional transient elastoplastic thermomechanical analysis. Comparative studies among conventional FE analysis and two different equivalent techniques, namely an inherent strain method and transient cooling-phase analysis to predict welding distortion, were carried out and summarized. Gannon et al. [68] investigated the influence of welding sequence on the distribution of residual stress and welding distortion generated when welding a flat-bar stiffener to a steel plate. The employed numerical simulation considered the element birth-and-death technique to represent the addition of weld metal to the workpiece during a sequential coupled thermal and structural analysis. The computed temperature profile and welding-induced residual stress and distortion were in good agreement with experimental measurements and analytical predictions. Then, a numerical study was carried out to discuss the distribution and magnitude of residual stress and welding distortion under four different welding sequences. Yan et al. [69] examined the welding distortion of a large aluminum-alloy stiffened sheet by friction stir welding (FSW) with numerical simulation. They showed that the predicted welding distortion pattern of the examined welded structure was convex in longitudinal direction and concave in transverse directions after FSW. The effect of stiffeners on residual welding distortion was also investigated. Owing to the adverse effect of welding distortion on products and fabrication processes, Wang et al. [70] focused on welding-induced buckling distortion with a global twisting shape mode. Both conventional thermal elastic–plastic FE analysis and an advanced elastic FE analysis with inherent deformation and interface elements were employed to investigate the twisting buckling behavior of the examined thin-plate stiffened structure under welding. Notably, the proposed elastic FE analysis provided results with very good agreement with experimental measurements while requiring much less computing time and resources than thermal elastic–plastic FE analysis for the same welding problem.

1.2.3.2 Elastic FE Computation with Welding Inherent Deformation

Given that elastic FE computation with shell-element model based on inherent strain/deformation has high computational efficiency and provides accurate results, it is usually employed for welding distortion investigation during ship structure fabrication. Luo et al. [71] pointed out that welding inherent strain is dominantly determined by the highest temperature reached during the thermal cycle and the self-constraint supported by surrounding materials. These authors conducted an elastic FE analysis to predict the welding distortion with the inherent strain as load. Wang et al. [72] established the relationship between inherent strain and welding configurations of typical welded joints with experimental results and thermal elastic–plastic FE computation. Then, inherent strains were applied as load to predict welding distortion of a large welded cylinder with multi-pass welding.

Later, Murakawa et al. [73] considered welding distortion as the result of elastic response of the examined welded structure with inelastic strains as load. Inelastic strains were totally defined as inherent strain and could be evaluated using a limited number of typical joints. Moreover, the inherent strains evaluated on each transverse cross-section were integrated to become welding inherent deformation, which is easier to store in databases for welding distortion prediction in the future. Luo et al. [74] predicted welding distortion by elastic FE analysis using inherent deformations and tendon force, which resulted from the integration of welding inherent strains. Numerical examples validated this approach. Luo et al. [75] created a database of welding inherent deformation from a series of thermal elastic–plastic FE computations of aluminum-alloy butt welded joint. Elastic FE analysis based on this database was carried out to predict the welding distortion of a large train coping with aluminum alloy plates. Deng and Murakawa [65] developed an elastic FE analysis to predict welding distortion accurately through the large-deformation theory. Thermal elastic–plastic FE computations were carried out to evaluate inherent deformations of various types of welded joints as a previous step, and then elastic FE computation was employed to predict welding distortion based on the evaluated inherent deformations. Murakawa et al. [76] summarized the concepts of inherent strain, inherent stress, inherent deformation, and inherent force during welding distortion prediction. Specifically, inherent strain became deformation when self-restraint was weak, and it became force when self-restraint was strong enough.

As applications of elastic FE computation for welding distortion prediction and mitigation during complex welded structure fabrication, Wang et al. [77] examined a spherical structure assembled with multiple thin-plates by elastic FE computation. The computed welding distortions were in good agreement with experimental measurements. Murakawa et al. [78] studied the welding distortion during the assembly process of a thin-plate structure with elastic FE computation. The impact of the assembly sequence, gap, and misalignment on the final dimensional accuracy was also investigated. Assembly processes such as cutting, forming, welding, and straightening during shipbuilding were all examined during the fabrication of a boxy structure, a ship block, a ship panel, and a super structure [79, 80]. The impact of gaps and misalignments during the fitting procedure on dimensional accuracy was also considered. In addition, Wang et al. [81] achieved welding distortion prediction in the fabrication of hatch coaming of a bulk cargo ship by elastic FE computation. Welding distortion was reduced by applying an improved assembly process with numerical analysis. Welding distortion in the fabrication of a cantilever beam component of jack-up drilling rig was also examined by elastic FE computation [82]. Mitigation techniques such as inverse distortion, welding sequence optimization, and mechanical constraint were applied to ensure dimensional accuracy.

1.3   Research Content

After the above introduction on research background and progress of fabrication processing mechanics through experimental and computational analysis, let us summarize how we addressed and examined the accurate fabrication of ship and offshore structures:

(1)   We focused on the fabrication processing of ship and offshore structures. Physical mechanisms of thermal transfer and mechanical response during cutting, plate bending, and welding were investigated and clarified.

(2)   Through an advanced measurement approach, the transient temperature, out-of-plane distortion, and residual stress after fabrication processing were comprehensively measured and evaluated.

(3)   To examine the thermal–mechanical response during fabrication processing, advanced thermal elastic–plastic FE computation was considered and elastic FE analysis based on inherent deformation was developed. Good agreement between the computed results and measurements was observed.

(4)   Cutting procedures with flame for thin plates of lightweight structures and thick plates of offshore structures were systematically examined. Distortion during thin-plate cutting and residual stress after thick-plate cutting were holistically evaluated.

(5)   Plate bending with singular and double curvatures was experimentally studied through high-frequency induction heating and computational approaches were carried out to represent the thermal–mechanical response. With good agreement and clarification of generation mechanism, an iterative approximation with the bisection method for processing parameter planning was achieved and validated.

(6)   Owing to the significant impact of welding distortion on dimensional accuracy, welding distortion prediction with computational welding mechanics was carried out during the fabrication of ship panels, major structures of a large container ship, and offshore structures. Practical methods of welding distortion mitigation were considered for accurate fabrication.

References

1. Di GB, Liu ZY, Hao L et al (2008) Production conditions and development trend of offshore platform steel. Mech Eng Mater 23(8):1–3 (in Chinese)
2. Zhou H, Jiang Z, Luo Y, Luo P (2014) Research on hull plate bending based on the technology of high-frequency induction. Shipbuild China 55(1):128–135
3. Heo SC, Seo YH, Ku TW, Kang BS (2010) A study on thick plate forming using flexible forming process and its application to a simply curved plate. Int J Adv Manuf Technol 51(1–4):103–115
4. Satoh K, Ueda Y, Fujimoto J (1979) Welding distortion and residual stresses. Sanpo Publications, Tokyo
5. Verhaeghe G (1999) Predictive formulae for weld distortion–a critical review. Abington Publishing, England Cambridge
6. Dean D, Ninshu Ma, Hidekazu M (2011) Finite element analysis of welding distortion in a large thin-plate panel structure. Trans JWRI 40(1):89–100
7. Dhingra AK, Murphy CL (2005) Numerical simulation of welding-induced distortion in thin-walled structures. Sci Technol Weld Join 10(5):528–536
8. Kim MJ (2005) 3D finite element analysis of evaporative laser cutting. Appl Math Model 29(10):938–954
9. Mohammadpour M, Razfar MR, Saffar RJ (2010) Numerical investigating the effect of machining parameters on residual stresses in orthogonal cutting. Simul Model Pract Theory 18(3):378–389
10. Yilbas BS, Arif AF, Abdul BJ (2010) Laser cutting of sharp edge: thermal stress analysis. Opt Lasers Eng 3:10–19
11. Maranh C, Paulo JD (2010) Finite element modeling of machining of AISI 316 steel: Numerical simulation and experimental validation. Simul Model Pract Theory 6:139–156
12. Eltawahni HA, Hagino M, Benyounis KY et al (2012) Effect of CO2 laser cutting process parameters on edge quality and operating cost of AISI316L. Opt Laser Technol 44:1068–1082
13. Wang XY, Meng QX, Kang RK et al (2010) Air melting ratio method Aluminum alloy thin plate cutting test research. China Laser 37(10):2648–2652 (in Chinese)
14. Xiao YP, Zhong JB , Miao SX et al (2010) S690 high-strength steel plate digital control flame cutting process. In: 2010 state steel structure science thesis book, vol 10, pp 565–568. (in Chinese)
15. Run PF, Wu YX, Liao K (2010) Simulation and analysis of cutting on the aluminum alloy thick plate residual stress distribution 41(6):2213–2219 (in Chinese)
16. Wu T, Kong XL, Wang X et al (2011) Cutting process research of the high-strength anti-wearing steel plate. New Technol New Process 6:85–87 (in Chinese)
17. Reddy JF (1971) Effects of high power laser radiation. Academic Press, NY
18. Dabby FW, Paek UC (1972) High intensity laser-induced vaporization and explosion of solid material. IEEE J Quant Elect QE-8 106–111
19. Bunting KA, Cornfield G (1975) Toward a general theory of cutting: a relationship between the incident power density and the cut speed. ASME J Heat Transf 97:116–121
20. Modest MF, Abakian H (1986) Evaporative cutting of a semi-infinite body with a moving CW laser. ASME J Heat Transf 108:597–601
21. Modest MF, Abakian H (1986) Heat conduction in a moving semi-infinite solid subjected to pulsed laser irradiation. ASME J Heat Transf 108:602–607
22. O'Neill W, Gabzdyl JT (2000) New developments in laser assisted oxygen cutting. Opt Lasers Eng 34:355–367
23. Gross MS, Black I, Muller WH (2003) Computer simulation of the processing of engineering materials with lasers–theory and first applications. J Phys D Appl Phys 36(2003):929–938
24. Gross MS, Black I, Muller WH (2005) 3-d simulation model for gas-assisted laser cutting. Lasers Eng 15:129–146
25. Gross MS (2006) On gas dynamic effects in the modeling of laser cutting processes. Appl Math Model 30:307–318

26. Mazumder J, Steen WM (1980) Heat transfer model for CW laser material processing. J Appl Phys 51:941–947
27. Kim MJ, Chen ZH, Majumdar P (1993) Finite element modelling of the laser cutting process. Comput Struct 49:231–241
28. Yu LM (1997) Three-dimensional finite element modeling of laser cutting. J Mater Process Technol 63:637–639
29. Kim MJ (2005) 3D finite element analysis of evaporative laser cutting. Appl Math Model 29:938–954
30. Powell J, Ivarson A, Magnusson C (1993) An energy balance for inert gas laser cutting. ICALEO, pp 12–20
31. Jang CD, Kim HK, Ha YS (2002) Prediction of plate bending by high-frequency induction heating. J Ship Prod 18(4):226–236
32. Shen H, Yao ZQ, Shi YJ et al (2006) Study on temperature field induced in high frequency induction heating. Acta Metall Sinica 19(3):190–196
33. Zhang X, Liu Y, Yang Y et al (2011) Technical parameter analysis of high-frequency induction heating applied to steel plate bending. J Ship Prod Des 27(27):99–110
34. Lee KS, Eom DH, Lee JH (2013) Deformation behavior of SS400 Thick plate by high-frequency induction heating based line heating. Met Mater Int 19(2):315–328
35. Lee KS, Kim SW, Eom DH (2015) Temperature distribution and bending behavior of thick metal plate by high frequency induction heating. Mater Res Innov 15(sup1):283–287
36. Bae KY, Yang YS, Hyun CM (2012) Analysis of triangle heating technique using high frequency induction heating in forming process of steel plate. J Precis Eng Manuf 13(4):539–545
37. Lee KS, Hwang B (2014) An approach to triangular induction heating in final precision forming of thick steel plates. J Mater Process Tech 214(4):1008–1017
38. Park DH, Jin HK, Park SS, Shin SB (2015) A study on the prediction of the angular distortion in line heating with high frequency induction heating. J Weld Join 33(1):80–86
39. Park J, Kim D, Mun SH et al (2016) Automated thermal forming of curved plates in shipbuilding: system development and validation. Int J Comput Integr Manuf 29(10):1128–1145
40. Zhang S, Liu C, Wang X et al (2018) Non-dimensional prediction of thermal forming behavior for the ship hull plate fabricated by induction heating. Ships Offshore Struct 2018:1–13
41. Wang JC, Huang WJ, Chang LC et al (2017) Elastic FE analysis on plate forming of pillow shape by line heating. Mar Technol 4:14–17
42. Bae KY, Yang YS, Hyun CM, Cho SH (2008) Derivation of simplified formulas to predict deformations of plate in steel forming process with induction heating. Int J Mach Tools Manuf 48(15):1646–1652
43. Zhang X (2011) Feasibility research on application of a high frequency induction heat to line heating technology. J Mar Sci Appl 10(4):456–464
44. Lee YH (2012) Prediction of plate bending by multi divisional analysis in induction heating. J Ship Res 56(3):146–153
45. Zhang X, Chen C, Li J, Liu Y (2017) The numerical study of steel plate forming by moveable induction heating considering the plate edge shrinkage. J Ship Prod Des 33(2):166–177
46. Dong H, Zhao Y, Yuan H (2018) Effect of coil width on deformed shape and processing efficiency during ship hull forming by induction heating. Appl Sci 8(9):1585–1601
47. Moshaiov A, Song H (1990) Near-and far-field approximation for analyzing flame heating and welding. J Therm Stress 13(1):1–19
48. Park JC, Jang KB, Cho SH, Jang TW (2005) Analysis of spot and line heating method for correcting thin plate deformation. In: Proceedings of the fifteenth international offshore and polar engineering conference, Seoul, pp 243–247
49. Wang JC, Rashed S, Murakawa H, Shibahara M (2011) Investigation of buckling deformation of thin plate welded structures. In: Proceedings of the twenty-first international offshore and polar engineering conference, Maui, pp 125–131
50. Wang JC, Rashed S, Murakawa H, Luo Y (2013) Numerical prediction and mitigation of out-of-plane welding distortion in ship panel structure by elastic FE analysis. Mar Struct 34:135–155

51. Blandon J, Sano M, Osawa N, Rashed S, Murakawa H (2014) Numerical study on mechanical characteristic of heat source for line heating. Prepr Ntnl Meeting JWS 404–405
52. Brown SB, Song H (1992) Implications of three-dimensional numerical simulations of welding of large structures. Weld J 71(2):55–62
53. Michaleris P, Debiccari A (1997) Prediction of welding distortion. Weld J 76(4):172–181
54. Jung GH, Tsai CL (2004) Plasticity-based distortion analysis for fillet welded thin-plate T-joints. Weld J 83(6):177–187
55. Jung GH (2007) A shell-element-based elastic analysis predicting welding induced distortion for ship panels. J Ship Res 51(2):128–136
56. Zhang L, Michaleris P, Marugabandhu P (2007) Evaluation of applied plastic strain methods for welding distortion prediction. J Manuf Sci Eng 129(6):1000–1010
57. Yang YP, Castner H, Kapustka N (2011) Development of distortion modeling methods for large welded structures. J Ship Prod Des 27(1):26–34
58. Michaleris P, DeBiccari A (1997) Prediction weld distortion. Weld J 76(4):172–180
59. Tsai CL, Park SC, Cheng WT (1999) Welding distortion of a thin-plate panel structure. Weld J 78(5):156–165
60. Vanli OA, Michaleris P (2001) Distortion analysis of welded stiffeners. J Ship Prod 17(4):226–240
61. Camilleri D, Comlekci T, Gray TGF (2006) Thermal distortion of stiffened plate due to fillet welds computational and experimental investigation. J Therm Stresses 29(2):111–137
62. Camilleri D, Mollicone P, Gray TGF (2006) Alternative simulation techniques for distortion of thin plate due to fillet welded stiffeners. Model Simul Mater Sci Eng 14:1307–1327
63. Jang CD, Lee CH, Ko DE (2002) Prediction of welding deforma- tion of stiffened panels. Proc Inst Mech Eng Part M J Eng Marit Environ 216:2133–2143
64. Huang TD, Dong PS, Decan L, Harwig D, Kumar R (2004) Fabrication and engineering technology for lightweight ship structure, part 1: distortion and residual stresses in panel fabrication. J Ship Prod 20(1):43–59
65. Deng D, Murakawa H, Liang W (2007) Numerical simulation of welding distortion in large structures. Comput Methods Appl Mech Eng 196(45–48):4613–4627
66. Biswas P, Mandal NR (2008) Welding distortion simulation of large stiffened plate panels. J Ship Prod 24(1):50–56
67. Biswas P, Mandal NR (2009) A comparative study of three different approaches of FE analysis for prediction of welding distortion of orthogonally stiffened plate panels. J Ship Prod 25(4):191–197
68. Gannon L, Liu Y, Pegg N, Smith M (2010) Effect of welding sequence on residual stress and distortion in flat-bar stiffened plates. Mar Struct 23(3):385–404
69. Yan D, Wu A, Silvanus J, Shi Q (2011) Predicting residual distortion of aluminum alloy stiffened sheet after friction stir welding by numerical simulation. Mater Des 32(4):2284–2291
70. Wang JC, Shibahara M, Zhang X, Murakawa H (2012) Investigation on twisting distortion of thin plate stiffened structure under welding. J Mater Process Technol 212(8):1705–1715
71. Luo Y, Murakawa H, Ueda Y (1997) Prediction of welding deformation and residual stress by elastic FEM based on inherent strain (first report): mechanism of inherent strain production. Trans JWRI 26(2):49–57
72. Wang JH, Luo H (2000) Prediction of welding deformation by FEM based on inherent strains. J Shanghai Jiaotong Univ (English Editorial Board) 5(2):83–87
73. Murakawa H, Luo Y, Ueda Y (1998) Inherent Strain as an interface between computational welding mechanics and its industrial application. Math Model Weld Phenom 4:597–619
74. Luo Y, Ishiyama M, Murakawa H (1999) Welding deformation of plates with longitudinal curvature. Trans JWRI 28(2):57–65
75. Luo Y, Deng D, Xie L, Murakawa H (2004) Prediction of deformation for large welded structures based on inherent strain. Trans JWRI 33(1):65–70
76. Murakawa H, Deng D, Ma N (2013) Concept of inherent strain, inherent stress, inherent deformation and inherent force for prediction of welding distortion and residual stress. In: Proceeding of the 1st international symposium on visualization in joining and welding science, 2013, Japan Osaka, pp 115–116

77. Wang JC, Ma X, Murakawa H, Teng BG, Yuan SJ (2011) Prediction and measurement of welding distortion of a spherical structure assembled form multi thin plates. Mater Des 32(10):4728–4737

78. Murakawa H, Deng D, Ma N, Wang JC (2012) Applications of inherent strain and interface element to simulation of welding deformation in thin plate structures. Comput Mater Sci 51(1):43–52

79. Murakawa H, Deng D, Rashed S, Sato S (2009) Prediction of distortion produced on welded structures during assembly using inherent deformation and interface element. Trans JWRI 38(2):63–69

80. Murakawa H, Okumoto Y, Rashed S, Sano M (2013) A practical method for prediction of distortion produced on large thin plate structures during welding assembly. Weld World 57(6):793–802. (November 2013)

81. Wang JC, Sano M, Rashed S, Murakawa H (2013) Reduction of welding distortion for an improved assembly process for hatch coaming production. J Ship Prod Des 29(4):1–9

82. Wang J, Zhao H, Zou J, Zhou H, Wu Z, Du S (2017) Welding distortion prediction with elastic FE analysis and mitigation practice in fabrication of cantilever beam component of jack-up drilling rig. Ocean Eng 130:25–39

Chapter 2
Fundamentals of FE Computation

To examine the physical behavior and generation mechanism during each fabrication processing, advanced FE computation, illustrated in Fig. 2.1, was developed and employed. It includes thermal elastic-plastic FE and elastic FE analyses based on the inherent deformation theory. In addition to this theory, other relevant theories such as finite strain theory, interface element, eigenvalue analysis are introduced next.

2.1 Non-linear Thermal Elastic-Plastic FE Computation

In thermal elastic-plastic FE analysis, two physical processes are considered, namely a thermal process and a mechanical process. Given that the thermal process has a decisive effect on the mechanical process whereas the mechanical process has only a small influence on the thermal process, thermal-mechanical behavior during welding is analyzed using uncoupled thermal-mechanical formulation. This uncoupled formulation considers the contribution of the transient temperature field to stress through thermal expansion and temperature-dependent mechanical properties.

The solution procedure comprises two steps:

(1) The transient temperature field of the heating process is calculated according to the heat transfer theory and high-temperature thermal properties, including thermal conductivity, thermal capacity, and density at different temperatures.

(2) The above transient temperature field is applied to mechanical analysis as thermal load. Then, residual strain, residual stress, and final deformation of the structure are obtained by using high-temperature mechanical properties, including the Young's modulus, Poisson's ratio, yield stress, and linear expansion coefficient at different temperatures.

The main thermal elastic-plastic computational model for transient thermal and mechanical analyses is described next.

© Science Press 2021
H. ZHOU and J. WANG, *FE Computation on Accuracy Fabrication of Ship and Offshore Structure Based on Processing Mechanics*,
https://doi.org/10.1007/978-981-16-4087-2_2

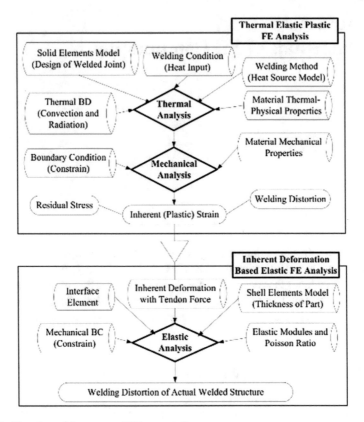

Fig. 2.1 Flowchart of the proposed FE computation

2.1.1   Transient Thermal Analysis

A welding employs a local high-intensity power source moving to the part. This source generates a sharp thermal profile in the weld pool, HAZ, and base metal. In a Lagrangian formulation, the 3D transient temperature is basically determined by solving the following partial differential equation for energy conservation on a domain defined by a FEM mesh:

$$
\begin{cases}
\dot{h} + \nabla \cdot q + Q = 0 \\
q = -k\nabla T \\
\dot{h} = \rho c \dfrac{dT}{dt}
\end{cases}
\tag{2.1}
$$

where h is the specific enthalpy and the superimposed dot denotes the derivative with respect to time, T is the temperature and ∇T is the temperature gradient, Q is the power per unit volume, i.e., the power density distribution, k is the thermal conductivity, c is the thermal capacity, and ρ is the density. These material properties are usually temperature-dependent.

In this analysis, the initial temperature is often assumed to be ambient temperature, and a body heat source with uniform power density (W/m^3: welding arc energy/volume of body heat source) is employed to model the heat source of the welding arc in a simple way.

Besides considering the moving heating source, heat losses due to convection and radiation are also taken into account in the FE model. Combining convection and radiation boundary conditions generates a boundary flux q(w/m^2) on all external surfaces. This flux is determined by joint heat convection and radiation as follows:

$$\begin{aligned}
q = q_c + q_r &= h(T - T_{ambient}) + \varepsilon C[(T + 273)^4 - (T_{ambient} + 273)^4] \\
&= [h + \varepsilon C[(T + 273)^2 + (T_{ambient} + 273)^2] \\
&\quad [(T + 273) + (T_{ambient} + 273)](T - T_{ambient})]
\end{aligned} \tag{2.2}$$

where q_c and q_r are the heat convection and radiation, respectively, h is the coefficient of heat convection, ε is the impassivity of the objective, which is equal to 1 for an ideal radiator, and C is the Stefan-Boltzmann constant.

The FEA domain of welding is dynamic, which means that it changes with the filler metal added during the welding pass per time step. After the welding pass is completed, the time-step length is exponentially increased until the sum of computed time reaches a prescribed maximum time, and the analysis halts.

2.1.2 Mechanical Analysis

Given the density ρ, the fourth-order elastic plastic tensor D as a 6×6 matrix, the body force b, and the Green-Lagrange strain ε, stress analysis solves the equation of momentum conservation at the end of each time step. This equation can also be written in the following form, in which inertial forces are neglected:

$$\begin{cases}
\nabla \cdot \sigma + b = 0 \\
\sigma = D\varepsilon \\
\varepsilon = \dfrac{\nabla_u + (\nabla_u)^T + (\nabla_u)^T \nabla_u}{2}
\end{cases} \tag{2.3}$$

The initial stress is often assumed to be zero. The system is solved using a time matching scheme with a prescribed time-step length during welding. Usually, an exponentially increasing time-step length after welding is stopped. In terms of boundary conditions, the part is free to deform but rigid-body motion is fixed as a constraint.

2.1.3 Fast Computation Techniques

Thermal elastic-plastic FE computation can completely represent the thermal and mechanical behaviors during a welding process. Computed results in good agreement with measurements can be obtained, but a large amount of computing resources and long computational time are required. Thus, advanced computational techniques to enhance computational efficiency are proposed.

2.1.3.1 Iterative Substructure Method

The most characteristic aspect in welding problems is that the region in strong non-linear state as a result of thermal elastic-plastic behavior is restricted to the vicinity of the welding heat source. The remaining area, which is much wider than the restricted region, is deemed in an elastic or weakly non-linear state. A second feature is that the non-linear area moves together with the heat source.

In usual FEM analysis, a non-linear problem against a whole region must be solved even in the case that a non-linear region is restricted into a narrow area. This results in a large-scale step-by-step non-linear computational problem. To prevent this inefficient computation, an iterative substructure method (ISM) was developed in which a whole computational area is divided into a weakly non-linear region and a strongly non-linear region [1]. As the strong non-linear region moves together with the heat source, the stiffness of the weakly non-linear region is updated only when necessary. To avoid the change of the stiffness matrix for this region, the original non-linear problem is divided into a linear problem and a non-linear problem, as shown in Fig. 2.2. The fictitious region corresponds to the strongly non-linear region at a certain instant of the welding process in the past.

Fig. 2.2 Image illustrating the proposed ISM

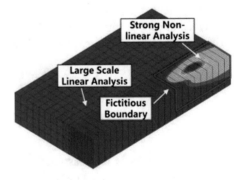

2.1.3.2 Parallel Computation

In a conventional TEP FE analysis, the program is coded and compiled in single-thread mode. Thus, it cannot exploit the advantage of multi-threads in high-performance servers for further computational efficiency. Parallel computation with Open Multi-Processing (OpenMP) technology was developed as a guided compilation processing scheme that can be implemented for multi-thread CPUs with shared-memory operation systems [2]. Then, effective computation, which was already accepted and employed for computational mechanics, can be achieved through multi-threads, as shown in Fig. 2.3.

With OpenMP technology, parallel computation is becoming much easier for researchers and scientists, and the program can be coded with many computer languages, such as C, C ++, or FORTRAN. Consider the multiplication of two ranked matrices multiplication given in Eq. (2.4) as an example. Parallel computation transforms these ranked matrix into vectors, and then vector multiplication is simultaneously carried out with different threads, as expressed in Eq. (2.5).

$$\begin{bmatrix} a_{11} & a_{12} \\ a_{21} & a_{22} \end{bmatrix} \begin{bmatrix} b_{11} & b_{12} \\ b_{21} & b_{22} \end{bmatrix} = \begin{bmatrix} c_{11} & c_{12} \\ c_{21} & c_{22} \end{bmatrix} \tag{2.4}$$

$$C_{11} = \begin{bmatrix} a_{11} \\ a_{12} \end{bmatrix}^T \begin{bmatrix} b_{11} \\ b_{21} \end{bmatrix} \quad C_{12} = \begin{bmatrix} a_{11} \\ a_{12} \end{bmatrix}^T \begin{bmatrix} b_{12} \\ b_{22} \end{bmatrix}$$

$$C_{21} = \begin{bmatrix} a_{21} \\ a_{22} \end{bmatrix}^T \begin{bmatrix} b_{11} \\ b_{21} \end{bmatrix} \quad C_{22} = \begin{bmatrix} a_{21} \\ a_{22} \end{bmatrix}^T \begin{bmatrix} b_{12} \\ b_{22} \end{bmatrix} \tag{2.5}$$

For computational welding mechanics, the reading and writing process are normally completed in single-thread mode, whereas the operation on large-range matrices is solved by OpenMP technology in multi-thread mode. Specifically, the heat flux matrix employed during thermal analysis and the mechanical stiffness matrix

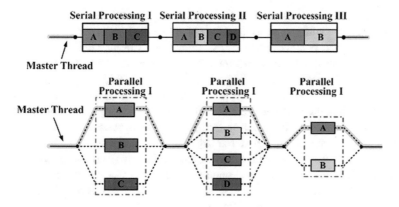

Fig. 2.3 Serial and parallel processing for multi-task computation

employed during mechanical analysis are divided into sub-matrices and arranged in different threads for parallel computation. Thus, many more threads are employed for one computation, and the computing time can be significantly reduced.

2.2   Theory of Inherent Strain and Deformation

If the inherent strain or inherent deformation is evaluated using a computational or an experimental approach, elastic FE analysis can be employed to predict welding distortion of welded structures, which is an ideal and practical computational approach to investigate welding distortion problems when fabricating large and complex welded structures. In this elastic FE analysis, the inherent strain/deformation is applied to the welding lines in shell-element models of welded structures. The interaction between the parts to be welded is represented by interface elements that are assumed to behave as a non-linear spring arranged between the parts to be welded. Bonding strength, gaps, and misalignments that arise during different assembly processes are considered by controlling the stiffness and deformation of the interface element.

During the induction heating process, the heated area expands and its magnitude is determined by the increased temperature, while compressive plastic strain is generated because of the constraint from the surrounding cold materials. When the compressive stress is higher than the yield stress of the materials, compressive plastic strain is generated. Conversely, tensile plastic strain is generated owing to the tensile stress in the cooling process. This partially decreases the compressive plastic strain. The remaining compressive plastic strain is retained to produce welding deformation and residual stress, that is, inherent strain [3].

2.2.1   Inherent Strain Theory

Based on a great deal of computed results from thermal elastic-plastic FE analysis and experimental observations, Ueda et al. [4, 5] concluded that residual stress and welding distortion are produced by the inherent strain ε^* during welding. This inherent strain mostly depends on interrelated parameters, such as the type of welded joint, material properties, plate thickness, and heat input. Taking a three-bar model as an example in which one side of the three bars is fixed and the other side is connected to move together, the inherent strain at each location is determined by the maximum temperature reached at that location during the heating process and the constraint supported by the surrounding material [6].

The total strain ε^{total} during the heating and cooling cycles of the welding process can be divided into the strain components given by Eq. (2.6) below, namely elastic strain $\varepsilon^{elastic}$, plastic strain $\varepsilon^{plastic}$, thermal strain $\varepsilon^{thermal}$, creep ε^{creep} strain, and the strain produced through phase transformation ε^{phase}:

$$\varepsilon^{total} = \varepsilon^{elastic} + \varepsilon^{thermal} + \varepsilon^{plastic} + \varepsilon^{phase} + \varepsilon^{creep} \qquad (2.6)$$

The total strain can be rearranged as a summation of the elastic strain and the inherent strain $\varepsilon^{inherent}$ that includes all the strain components except the elastic strain. In other words, the inherent strain $\varepsilon^{inherent}$ is defined as a summation of plastic strain, thermal strain, creep strain, and the strain caused by the phase transformation, as given by Eq. (2.7) below. Especially, for welded joints made of carbon steel, the inherent strain can be represented by the plastic strain because the strain induced by creep and solid-state phase transformation is much smaller, and the thermal strain disappears when the temperature cools down to the ambient temperature or initial temperature. For welding of phase-transformation material, the phase-transformation strain is generated when the solid-state phase transformation occurs during a cooling process with high cooling rate. This type of strain influences the magnitude of inherent strain, eventually influencing the residual stress and welding distortion as well.

$$\varepsilon - \varepsilon^{total} = \varepsilon^{thermal} + \varepsilon^{plastic} + \varepsilon^{phase} + \varepsilon^{creep} = \varepsilon^{inherent} = \varepsilon^* \qquad (2.7)$$

2.2.2 Inherent Deformation Theory

Owing to the concentration characteristic of inherent strain and the requirement of fine enough FE mesh near the welding line to apply inherent strain, the distribution of inherent strain is difficult to be directly employed for predicting the residual stress and the welding distortion with an elastic analysis.

The inherent deformation theory is based on the assumption that a welded joint in a plate structure has an inherent amount of deformation on every cross-section normal to the welding line during the welding process. This theory can be employed to replace the distribution of inherent strain for each cross-section with one value for elastic FE analysis. Given that the displacement or the deformation results from the integration of strain, inherent deformation, which results from the integration of inherent strain, can be used to predict welding distortion without a significant loss of accuracy.

Similar to the inherent strain, the inherent deformation mostly depends on inter-related parameters, such as configuration, material properties, plate thickness, and heat input. The influence of the length and the width of a welded joint is small if the size of the plate is large enough [7]. When the effect of edges is ignored, the components of inherent deformation can be approximated as constant values along the welding line. These constant values are introduced into the elastic model as loads (forces and displacements) to predict welding distortion. Computed results are in good agreement with experimental measurements [8].

The inherent deformation can be evaluated as the integration of the longitudinal inherent strain in the welding direction and the transverse inherent strain in the transverse direction distributed on the cross-section normal to the welding line according to the following equations:

$$
\begin{cases}
\delta_x^* = \frac{1}{h} \iint \varepsilon_x^* dydz \quad \theta_x^* = \frac{12}{h^3} \iint \left(z - \frac{h}{2}\right) \varepsilon_x^* dydz \\
\delta_y^* = \frac{1}{h} \iint \varepsilon_y^* dydz \quad \theta_y^* = \frac{12}{h^3} \iint \left(z - \frac{h}{2}\right) \varepsilon_y^* dydz
\end{cases}
\tag{2.8}
$$

where δ_x^* and δ_y^* respectively are the inherent deformations in the longitudinal and transverse directions, and θ_x^* and θ_y^* respectively are the inherent bending deformations in the longitudinal and transverse directions; h is the thickness of the welded joint, and x,y,z are the welding, transverse and thickness directions, respectively.

The longitudinal inherent deformation has a different nature from other inherent deformation components. This is due to the constraint provided by the surrounding material, which prevents free shrinkage along the welding line. Therefore, the longitudinal inherent deformation appears as a tensile force acting in the welding line, which is referred to as the tendon force.

Furthermore, the relation between the longitudinal inherent deformation δ_L^* and the tendon force F_{tendon} (longitudinal inherent shrinkage force) [9] can be derived from their definitions expressed as follows:

$$
F_{tendon} = \int E \times \varepsilon_L^* dA = E \times h \times \frac{1}{h} \int_L \varepsilon_L^* dA = Eh\delta_L^*
\tag{2.9}
$$

where ε_L^* and δ_L^* respectively are the inherent strain and inherent deformation in the longitudinal direction, and h is the thickness of the welded joint.

2.3 Interface Element

If all parts to be assembled by welding have no geometrical errors, they are fully fitted by tack welding with sufficient stiffness. Moreover, if they are welded simultaneously, the final distortion of the structure after welding assembly is solely determined by welding inherent deformation.

However, in actual assembly processes of large-scale welded structures, the parts are sequentially assembled by repeated fitting, tack welding, and welding. In this situation, owing to welding distortion, the already assembled members no longer have the designed geometry or dimensions. This leads to gaps and misalignments in the fitting stage, which can also be produced by geometrical errors due to cutting and formation. If the gap or misalignment between two parts to be welded exceeds the tolerable limit, they are corrected during the fitting process before welding. In this case, the final distortion of the whole structure is influenced by gaps and misalignments and their correction during fitting. Therefore, gaps and misalignments should be considered to predict and control the distortion of a large structure during the assembly process. The evolution of the mechanical interaction between parts to be assembled through fitting, tack welding, and fully welded states can be conveniently described by the interface element [10, 11], which is defined as a non-linear spring arranged between the parts to be welded, as shown in Fig. 2.4. The relative displacements or discontinuities of the deformation across the interface element in the welding direction, normal direction, and transverse direction are denoted as u_L, u_N, and u_T, respectively. The rotation around the welding direction is denoted by u_θ. The forces and moment associated with these displacements are denoted by f_L, f_N, f_T, and f_θ, respectively. The relations between the displacements and forces are also illustrated in Fig. 2.4. The mechanical properties of the non-linear spring are controlled by the stiffness K, the maximum force f_{max}, and the gap u_G. When the members are free, $K = 0$. In the fitting stage, the value of stiffness K for each direction is set to appropriate values according to the type of tack welding or fixture. The gap in the fitting stage can be controlled by the values of K and f_{max}. When the joint is fully connected after welding, the stiffness is set to a large enough value.

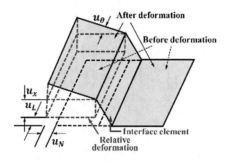

(a) Definition of displacement

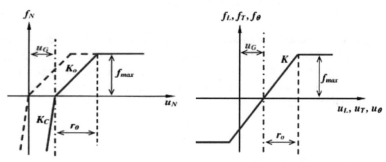

(b) Mechanical properties of interface element (normal direction).

(c) Mechanical properties of interface in interface element (transverse and rotational directions)

Fig. 2.4 Application of interface element in elastic FE analysis for welding

2.4 Elastic Buckling Theory

Welding induced buckling considered as particular out-of-plane welding distortion not only influences the dimensional tolerance but also increases the fabrication cost and schedule due to expensive consumption when mitigated. Therefore, it is better to predict the behavior of welding induced buckling before actual fabrication, and even evaluate the critical buckling condition to avoid its occurrence during advanced manufacturing of lightweight structures.

2.4.1 Finite Strain Theory

The equation relating the strain and displacement is essential to describe the buckling behavior. If a small deformation is assumed, the strains are expressed as a linear function of displacements. When large deformation is considered, Green-Lagrange strain as expressed in Eq. (2.10) below is employed. The strains present a non-linear relationship with respect to the displacements:

$$
\begin{cases}
\varepsilon_x = \frac{\partial u}{\partial x} + \frac{1}{2}\left\{ \left(\frac{\partial u}{\partial x}\right)^2 + \left(\frac{\partial v}{\partial x}\right)^2 + \left(\frac{\partial w}{\partial x}\right)^2 \right\} \\
\varepsilon_y = \frac{\partial u}{\partial y} + \frac{1}{2}\left\{ \left(\frac{\partial u}{\partial y}\right)^2 + \left(\frac{\partial v}{\partial y}\right)^2 + \left(\frac{\partial w}{\partial y}\right)^2 \right\} \\
\varepsilon_z = \frac{\partial u}{\partial z} + \frac{1}{2}\left\{ \left(\frac{\partial u}{\partial z}\right)^2 + \left(\frac{\partial v}{\partial z}\right)^2 + \left(\frac{\partial w}{\partial z}\right)^2 \right\} \\
\gamma_{xy} = \frac{\partial u}{\partial y} + \frac{\partial v}{\partial x}\left\{ \left(\frac{\partial u}{\partial x}\right)\left(\frac{\partial v}{\partial y}\right) + \left(\frac{\partial v}{\partial x}\right)\left(\frac{\partial v}{\partial y}\right) + \left(\frac{\partial w}{\partial x}\right)\left(\frac{\partial w}{\partial y}\right) \right\} \\
\gamma_{yz} = \frac{\partial v}{\partial z} + \frac{\partial w}{\partial y}\left\{ \left(\frac{\partial u}{\partial y}\right)\left(\frac{\partial v}{\partial z}\right) + \left(\frac{\partial v}{\partial y}\right)\left(\frac{\partial v}{\partial z}\right) + \left(\frac{\partial w}{\partial y}\right)\left(\frac{\partial w}{\partial z}\right) \right\} \\
\gamma_{zx} = \frac{\partial v}{\partial x} + \frac{\partial w}{\partial z}\left\{ \left(\frac{\partial u}{\partial z}\right)\left(\frac{\partial v}{\partial x}\right) + \left(\frac{\partial v}{\partial z}\right)\left(\frac{\partial v}{\partial x}\right) + \left(\frac{\partial w}{\partial z}\right)\left(\frac{\partial w}{\partial x}\right) \right\}
\end{cases}
\tag{2.10}
$$

where ε_x, ε_y, and ε_z are Green-Lagrange strains in x, y, and z directions respectively; γ_{xy}, γ_{yz}, and γ_{zx} are the shear strains on the $x-y, y-z$, and $z-x$ planes, respectively; and u, v, and w are the displacements in x, y, and z directions, respectively. In Eq. (2.10), the first-order terms express linear response whereas the second-order terms are essential to express non-linear behavior, such as buckling.

2.4.2 Eigenvalue Analysis

Eigenvalue analysis is employed to predict the buckling force of a structure assumed as an ideal linear elastic body. In classical eigenvalue analysis, eigenvalues are computed with regard to the applied compressive force and constraints of a given system. For a basic structural configuration, each applied combination of forces has a minimum critical buckling value at which the structure buckles and a corresponding buckling mode.

Eigenvalue analysis also produces critical buckling values and the associated buckling modes. Theoretically, there are values and modes equal in number to the number of degrees of freedom in the considered system.

Basic variables such as the displacement $u_t + \Delta t$, the strain $\varepsilon_t + \Delta t$, and the stress $\delta_t + \Delta t$ at time $t + \Delta t$ can be decomposed into their values at time t and their increments as follows:

$$u_t + \Delta t = u_t + \Delta u$$
$$\varepsilon_t + \Delta t = \varepsilon_t + \Delta^1 \varepsilon + \Delta^2 \varepsilon$$
$$\sigma_t + \Delta t = D\varepsilon_t + \Delta t = \sigma_t + D\Delta^1 \varepsilon + D\Delta^2 \varepsilon \qquad (2.11)$$

where D is the elastic matrix (stress-strain matrix), and $\Delta^1 \varepsilon$, $\Delta^2 \varepsilon$ denote the first-order (linear) and second-order (non-linear) terms of strain increment, respectively.

When Eq. (2.10) is used, and taking ε_x as an example, we obtain the following equation:

$$\varepsilon_x(t + \Delta t) = \frac{\partial u_t + \Delta t}{\partial x} + \frac{1}{2}\left(\frac{\partial u + \Delta t}{\partial x}\right)^2 + \frac{1}{2}\left(\frac{\partial v_t + \Delta t}{\partial x}\right)^2 + \frac{1}{2}\left(\frac{\partial w_t}{\partial x}\right)^2$$
$$= \varepsilon_x(t) + \Delta^1 \varepsilon_x + \Delta^2 \varepsilon_x \qquad (2.12)$$

where

$$\varepsilon_x(t) = \frac{\partial u_t}{\partial x} + \frac{1}{2}\left(\frac{\partial u_t}{\partial x}\right)^2 + \frac{1}{2}\left(\frac{\partial v_t}{\partial x}\right)^2 + \frac{1}{2}\left(\frac{\partial w_t}{\partial x}\right)^2$$

$$\Delta^1 \varepsilon_x = \frac{\partial \Delta u}{\partial x} + \frac{\partial u_t}{\partial x}\frac{\partial \Delta u}{\partial x} + \frac{\partial v_t}{\partial x}\frac{\partial \Delta v}{\partial x} + \frac{\partial w_t}{\partial x}\frac{\partial \Delta w}{\partial x}$$

$$\Delta^2 \varepsilon_x = \frac{1}{2}\left(\frac{\partial \Delta u}{\partial x}\right)^2 + \frac{1}{2}\left(\frac{\partial \Delta v}{\partial x}\right)^2 + \frac{1}{2}\left(\frac{\partial \Delta w}{\partial x}\right)^2$$

To derive the governing equation for eigenvalue analysis, the minimum potential energy theorem is employed. The total energy of the system at time t and $t + \Delta t$ are given by Eqs. (2.13) and (2.14), respectively:

$$\pi(u_t) = \int \frac{1}{2}\varepsilon_t^T D\varepsilon dv - \int f_t u_t ds \qquad (2.13)$$

$$\pi(u_t + \Delta_t) = \int \frac{1}{2}(\varepsilon_t + \Delta^1 \varepsilon + \Delta^2 \varepsilon)^T D(\varepsilon_t + \Delta^1 \varepsilon + \Delta^2 \varepsilon)dv$$

$$- \int (f_t + \Delta f)(u_t + \Delta u)ds = \pi(u_t) + \Delta\pi(\Delta u) \qquad (2.14)$$

Given that the potential energy at time t is already known and fixed, the condition for the potential energy $\pi(u_t + \Delta t)$ to be minimum is equivalent to the condition for its increment $\Delta\pi(\Delta u)$, given by Eq. (2.14), in which the higher-order terms are neglected. Equation (2.14) can be rewritten in matrix form as follows:

$$\Delta\pi(\Delta u) = \pi(u_t + \Delta u) - \pi(u_t)$$
$$= \int \frac{1}{2}\left\{(\Delta^1\varepsilon)^T D(\Delta^1\varepsilon) + 2D\varepsilon_t\Delta^1\varepsilon\right\}dv - \int (f_t\Delta u + \Delta f u_t)ds$$
$$= \int \frac{1}{2}\left\{(\Delta^1\varepsilon)^T D(\Delta^1\varepsilon) + 2\sigma_t\Delta^2\varepsilon\right\}dv - \int \Delta f u_t ds + \int \sigma_t\Delta^1\varepsilon dv - \int f_t\Delta u ds$$
$$(2.15)$$

Equation (2.15) can be rewritten in the following matrix form:

$$\Delta\pi(\Delta u) = \frac{1}{2}\{\Delta u\}^T[K_1(u_t)]\{\Delta u\}$$
$$+ \frac{1}{2}\{\Delta u\}^T[K_2(\sigma_t)]\{\Delta u\} - \{\Delta f\}^T\{\Delta u\} + \{F\}^T\{\Delta u\} - \{f_t\}\{\Delta u\}$$
$$(2.16)$$

where

$$\frac{1}{2}\{\Delta u\}^T[K_1(u_t)]\{\Delta u\} = \int \frac{1}{2}(\Delta^1\varepsilon)^T D(\Delta^1\varepsilon)dv$$
$$\frac{1}{2}\{\Delta u\}^T[K_2(u_t)]\{\Delta u\} = \int D\varepsilon\Delta^2\varepsilon dv = \int \sigma_t\Delta^2\varepsilon dv$$
$$\{F\}^T\{\Delta u\} = \int D\varepsilon_t\Delta^1\varepsilon dv \cdots \{\Delta f\}^T\{\Delta u\} = \int \Delta f\Delta u_t dv \cdots \{f_t\}^T\{\Delta u\} = \int f_t\Delta u dv$$

From the condition for minimum value of $\Delta\pi(\Delta u)$, the following equation is derived:

$$\frac{\partial \Delta\pi}{\partial \Delta u} = [K_1(u_t)]\{\Delta u\} + [K_2\{\sigma_t\}]\{\Delta u\} + \{F\} - \{f_t\} - \{\Delta f\} = 0 \qquad (2.17)$$

According to the equilibrium of the system at time t, we can write the following expression:

$$\{F\} - \{f_t\} = 0 \qquad (2.18)$$

Because of the fact that buckling occurs without increase of external forces, the following equation is satisfied:

$$\{\Delta f\} = 0 \tag{2.19}$$

If the system buckles under an internal force $\lambda\{\sigma_t\}$, Eq. (2.19) is reduced to the following eigenvalue problem:

$$[K_1(u_t)]\{\Delta u\} + \lambda[K_2\{\sigma_t\}]\{\Delta u\} = 0 \tag{2.20}$$

From Eq. (2.20), it is clear that buckling is an eigenvalue problem. In case of welding, the stress σ_t is the stress produced by the inherent deformation associated with the welding whereas the parameter λ is the eigenvalue to be determined. When the stress becomes equal to $\lambda\sigma_t$, the structure buckles. This means that the structure has already buckled when λ is less than 1.0.

2.5 Tendon Force and Its Evaluation

With the wide application of welding in shipbuilding, welding distortion, in particular welding induced buckling in thin-plate welded structures, has become a critical welding problem. All types of welding distortion are the results of non-uniform expansion/contraction and the associated plastic strain produced in the weld and surrounding base material, which are caused by the heating and cooling cycles during the welding process. More specifically, due to the constraint supported by the surrounding colder material on the expansion and sub-sequential contraction of the heating material, compressive plastic strain is produced. Then, tensile residual stress appears in and around the welding line, combined with compressive residual stress simultaneously generated in locations of the parts away from the welding region.

Generally speaking, the welding distortion of plate welded structures can be classified into in-plane and out-of-plane distortions. Welding shrinkage is the typical in-plane type; out-of-plane welding distortion is primarily divided into two different types: bending and buckling distortions. For a single welded joint, longitudinal shrinkage and welding induced buckling is dominantly determined by the tendon force, which is also considered as a longitudinal inherent shrinkage force. This force has a strong tensile nature and small acting area resembling a tendon.

In 1980, White et al. [12] first introduced the concept of tendon force and proposed a formula based on an experimental measurement described as follows:

$$F_{tendon} = 0.2Q_{net}(Q_{net} = \eta Q) \tag{2.21}$$

where F_{tendon} is expressed in kN, Q_{net} (J/mm) is the net heat input per unit length, and η is the heat efficiency.

Sato et al. [13] proposed a relation between the tendon force and the heat input, as shown in Eq. (2.22). Terasaki et al. [14] concluded that the relation between tendon force and heat input is influenced by the thickness of the welded plate. For thin plates, Eq. (2.23) can be used to evaluate the tendon force. The relation expressed by Eq. (2.24) is suitable for thick plates. Luo et al. [15] presented the relation shown in Eq. (2.25), which utilizes thermal elastic-plastic FE analysis. They pointed out that the tendon force is not produced until the heat input exceeds a critical value.

$$F_{tendon} = 0.107 Q_{net} \tag{2.22}$$

$$F_{tendon} = 0.215 Q_{net} \tag{2.23}$$

$$F_{tendon} = 0.159 Q_{net} \tag{2.24}$$

$$F_{tendon} = 0.172(Q_{net} - 28.56) \tag{2.25}$$

By employing the tendon force and other components of inherent deformation, elastic FE computation is used to predict welding distortion, in particular welding induced buckling. Luo et al. [16] first analyzed welding deformation of plates with longitudinal curvature using thermal elastic-plastic FE analysis. Luo suggested that given that both welding distortion and residual stress are produced by inherent strain, then the integration of the inherent strain provides not only the inherent deformations but also the tendon force. A method to predict the welding deformation by elastic FE analysis using inherent deformations and the tendon force was proposed and its validation was demonstrated through numerical examples. Tajima et al. [17] pointed out that welding longitudinal shrinkage (tendon force) produces compressive stress in the surrounding plate fields. This sometimes causes these plate fields to buckle. To avoid buckling, a series of thermal elastic-plastic FE analyses were carried out to predict welding tendon forces and transverse shrinkage/bending when utilizing continuous and intermittent welding with different welding specifications. A cross-stiffened panel of a car deck in a car carrier was investigated. Bi-directional residual stresses were then evaluated for different welding patterns. The effectiveness of welding patterns (continuous, parallel, and zigzag intermittent welding) for reducing the welding residual stress and preventing buckling was quantified.

Wang et al. [18] focused on the residual buckling distortion in a test specimen with a thickness of 2.28 mm under bead-on-plate welding. Eigenvalue and elastic FE analyses based on the inherent deformation theory were carried out to investigate the generation mechanism of residual buckling in bead-on-plate welded joints. The tendon force (longitudinal inherent shrinkage) is the dominant reason to produce buckling, and the disturbance (initial deflection or inherent bending) triggers buckling but does not influence the buckling mode. In addition, Wang et al. [19] focused on welding induced buckling for welded structures assembled by thin plates. Tendon force produced along the welding line due to welding was introduced to explain the mechanism of buckling behavior. In addition, thermal elastic-plastic FE analysis was carried out to consider the influence of the heat input on the magnitude of tendon force when employing the small- and large-deformation theories. Eigenvalue analysis showed that the tendon force is the dominant source of welding induced buckling. However, the buckling was not represented in the computed results of elastic FE analysis because of the absence of disturbance that initiates the buckling. These authors also pointed out that the eigenvalue analysis reports when the buckling will happen and that the magnitude of buckling distortion depends not only on the critical compressive force but also on the initial disturbance or deflection. Deng et al. [20] proposed to conduct elastic FE analysis based on the inherent deformation method to predict the welding distortion accurately during the assembly process considering both local shrinkage and root gap. To obtain the magnitude of all the components of inherent deformation for different welded joints in the examined welded structure, thermal elastic-plastic FE computation was employed. Specifically, the tendon force in the fillet welded joint was studied. These authors concluded that after the first welding, the coefficient between the tendon force and the heat input was approximately 196.0, which is almost the same value produced by the formula proposed by White. However, after the second welding, this coefficient drastically decreased to 126.0. The reason for this was that the plastic zone generated after the first welding was affected by the second welding.

2.5.1 Tendon Force Evaluation with Theoretical Analysis

Considering an ideal welding case in which the arc heating penetrates a wide enough plate with a high moving speed, a simplified mechanical model that is fixed with the strong constraint supported by the surrounding material is proposed to represent welding distortion behavior in the longitudinal direction. The resulting theoretical solution shows that the longitudinal inherent strain and its corresponding magnitude [6] on the cross-section normal to the welding line can be derived for the three regions according to the maximum temperature T_{max} as expressed in the following expression:

$$\begin{cases} T_{max} \leq T_Y & \varepsilon_L^* = 0 \\ T_Y > T_{max} < 2T_Y & \varepsilon_L^* = -\alpha(T_{max} - T_Y) \\ 2T_Y \leq T_{max} & \varepsilon_L^* = -\alpha T_Y \end{cases} \tag{2.26}$$

where T_Y is the yield temperature and α is the linear thermal expansion coefficient (unit: 1/°C).

Furthermore, it is well-known that the relation between the tendon force F_{tendon} and the longitudinal inherent strain ε_L^* in an infinitesimal area dA is given by

Eq. (2.27) below. Concerning Eq. (2.28) next, it is employed to evaluate the tendon force as an integration of the longitudinal inherent strain over the cross-section normal to the welding direction:

$$dF_{tendon} = E\varepsilon_L^* dA \tag{2.27}$$

$$F_{tendon} = E \times \int\int_{-\infty}^{\infty} \varepsilon_L^* dy dz = 2Eh \times \int_0^{\infty} \varepsilon_L^* dy \tag{2.28}$$

According to the theoretical solution of thermal conduction in the transverse direction, the positions y_1 and y_2 that reach the maximum temperature $T_Y(T_1)$ and $2T_Y(T_2)$ respectively can be calculated as follows:

$$\begin{cases} y_1 = \dfrac{Q_{net}}{\rho c h \sqrt{2\pi e}} \dfrac{\alpha E}{\sigma_Y} \\[2ex] y_2 = \dfrac{Q_{net}}{2\rho c h \sqrt{2\pi e}} \dfrac{\alpha E}{\sigma_Y} \end{cases} \tag{2.29}$$

where Q_{net} is the net heat input by arc welding, and σ_Y is the yield stress of the examined metal.

Integrating the inherent strain on the whole cross section, the tendon force (longitudinal inherent shrinkage force) produced by welding can be evaluated as follows:

$$\begin{aligned} F_{tendon} &= 2h\left(\int_0^{y_2} E \times (-\alpha T_Y)dy + \int_{y_2}^{y_1} -\alpha(T_{max} - T_Y) \times E dy + \int_{y_1}^{+\infty} E \times 0 dy \right) \\ &= 2h\left(\int_0^{y_2} E \times (-\alpha T_Y)dy + \int_{y_2}^{y_1} \alpha T_Y \times E dy + \int_{y_2}^{y_1} -\alpha T_{max} \times E dy \right) \\ &= 2h\left(-\alpha T_Y E(y_2 - 0) + \alpha T_Y E(y_1 - y_2) + (-\alpha E)\frac{Q_{net}}{\rho c h \sqrt{2\pi e}} \ln\left(\frac{y_1}{y_2}\right) \right) \\ &= (-\alpha E)\frac{2Q_{net}}{\rho c \sqrt{2\pi e}} \ln(2) = -0.335 \times \frac{\alpha E}{\rho c} \times Q_{net} \end{aligned} \tag{2.30}$$

When the material properties of carbon steel at room temperature are used, the tendon force can be rearranged as given by Eq. (2.31) below. The negative sign in this equation means that the tendon force is active as a compressive force:

$$F_{tendon} = -0.335 \times \frac{E\alpha}{\rho c}\eta Q = -0.235 Q_{net} \tag{2.31}$$

where η is the welding heat efficiency, E (2.1×10^6 MPa) is the Young's modulus, α ($1.2 \times 10^{-5}/°C$) is the coefficient of line expansion, ρ (7800 kg/m^3) is the density, and c (4.6×10^2 J/(kg·°C)) is the specific heat of material.

2.5.2 Tendon Force Evaluation by Computed Results

To predict welding distortion using elastic FE analysis, the inherent deformation including tendon force should be previously evaluated. Owing to the importance of the tendon force on longitudinal shrinkage and welding induced buckling, three practical approaches to evaluate its magnitude, namely evaluation using measured longitudinal displacement, a theoretical formula, and integration of inherent strain or inherent stress, are presented next.

2.5.2.1 Direct Approach of Evaluation with Displacement

Results of thermal elastic-plastic FE analysis show the distribution of longitudinal displacement along the welding line. The curve can be divided into a middle part and two ends. In the middle part, the curve is almost linear with a constant slope [22]. From this slope, the longitudinal inherent deformation can be evaluated as described next.

Given that the slope of the longitudinal displacement in Fig. 2.5 is the longitudinal strain ε_L in the welding direction ($\varepsilon_L = \partial u/\partial x$), it can be related to the tendon force F_{tendon} and the longitudinal inherent deformation δ_L^* through the following equation:

$$\varepsilon_L = \frac{F_{tendon}}{AE} = \frac{Eh\delta_L^*}{BhE} = \frac{\delta_L^*}{B} \qquad (2.32)$$

or

$$\delta_L^* = B\varepsilon_L \qquad (2.33)$$

Fig. 2.5 Distribution of longitudinal displacement along the welding line

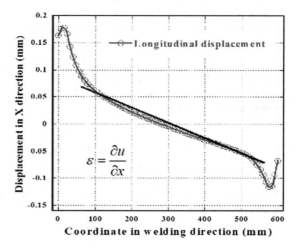

where δ_L^* is the inherent deformation in x direction of a length equal to B, and B is the width of the welded joint.

Equation (2.33) means that the longitudinal inherent deformation δ_L^* can be evaluated as the longitudinal displacement between two points separated in x direction by a length equal to the plate width B. If the specimen or the model is long enough, Eq. (2.33) can be used to evaluate the longitudinal inherent deformation both in experiments and in thermal elastic plastic FE analysis as follows:

$$F_{tendon} = Eh\delta_L^* = 49.02KN \tag{2.34}$$

2.5.2.2   Integration of Inherent Strain

According to the definition of inherent deformation given by Eq. (2.8), the longitudinal inherent deformation can be evaluated by integrating the longitudinal inherent strain. By utilizing the relationship between the tendon force and longitudinal inherent deformation, the later can be evaluated by integration of the longitudinal inherent stress, which can be converted from longitudinal residual stress.

As discussed above, in carbon steel, the plastic strain is the dominant component of the inherent strain. Plastic strain can be obtained by thermal elastic-plastic FE analysis. Then, the longitudinal inherent deformation δ_L^* may be evaluated by using the longitudinal inhere strain ε_L^* on a transverse cross-section at the middle of the welded joint:

$$\delta_L^* = \frac{1}{h} \iint \varepsilon_L^* dydz \tag{2.35}$$

If the same integrating process is carried out for each transverse cross-section, the distribution of the longitudinal inherent deformation along the welding line can be obtained. Note from this figure that the distribution of the longitudinal inherent deformation in the middle region of the welding line is almost constant.

For the special case of a very thin plate, the distribution of longitudinal inherent strain is nearly uniform along the thickness direction. Thus, the thickness of a welded joint is approximately equal to the thickness of a welded plate, and Eq. (2.35) can be simplified as follows:

$$\delta_L^* = \int \varepsilon_L^* dy \tag{2.36}$$

where ε_L^* is the longitudinal inherent strain considered to be uniform along the thickness direction.

2.6 Conclusions

Elementary concepts and theories as well as computational approaches for mechanical response investigation of fabrication processing during the construction of ship and offshore structures were summarized. In particular, elastic FE computation, which is considered as a practical and efficient approach for bending distortion prediction and mitigation during large-structure fabrication, was systematically introduced. Welding inherent deformation evaluation and welding induced buckling were also clarified.

References

1. Nishikawa H, Serizawa H, Murakawa H (2007) Actual application of FEM to analysis of large scale mechanical problems in welding. Sci Technol Weld Join 12(2):147–152
2. Chandra R, Dagum L, Kohr D, Maydan D, Mcdonald J, Menon R (2001) Parallel programming in OpenMP. Academic Press
3. Ueda Y, Murakawa H, Ma N (2012) Welding deformation and residual stress prevention. Elsevier Publishing
4. Ueda Y, Murakawa H, Nakacho K et al (1995) Establishment of computational welding mechanics. Trans Join Weld Res Inst 24(2):73–86
5. Luo Y, Murakawa H, Ueda Y (1997) Prediction of welding deformation and residual stress by elastic FEM based on inherent strain (first report): mechanism of inherent strain production. Trans Join Weld Res Inst 26(2):49–57
6. Ueda Y, Murakawa H, Ma N (2012) Welding deformation and residual stress prevention. Butterworth-Heinemann Publications
7. Vega A, Rashed S, Serizawa S, Murakawa H (2007) Inflfluential factors affecting inherent deformation during plate forming by line heating (report 1): the effect of plate size and edge effect. Trans JWRI 36(2):57–64
8. Wang JC, Ma X, Murakawa H, Teng BG, Yuan SJ (2011) Prediction and measurement of welding distortion of a spherical structure assembled form multi thin plates. Mater Des 32(10):4728–4737
9. Wang JC, Matsuo Y, Murakawa H (2012) Numerical study on longitudinal inherent deformation and inherent force Produced in bead on plate welding. Preprints of the Natl Meet Jpn Weld Soc 91:162–163
10. Murakawa H, Deng D, Rashed S, Sato S (2009) Prediction of distortion produced on welded structures during assembly using inherent deformation and interface element. Trans JWRI 38(2):63–69
11. Dean D, Ma N, Murakawa H (2011) Finite element analysis of welding distortion in a large thin-plate panel structure. Trans JWRI 40(1):89e100
12. White JD, Leggatt RH, Dwight JB (1980) Weld shrinkage prediction. Weld Metal Fabr 11:587–596
13. Satoh K, Ueda Y, Fujimoto J (1979) Welding distortion and residual stresses. Sanpo Publications
14. Terasaki T, Nakatani M, Ishimura T (2000) Study of tendon force generating in welded joint. J Jpn Weld Soc 18(3):479–486
15. Luo Y, Lu HY, Xie L et al (2004) Concept and evaluated method of tendon force. Mar Technol 4:35–37
16. Luo Y, Ishiyama M, Murakawa H (1999) Welding deformation of plates with longitudinal curvature. Trans Join Weld Res Inst 28(2):57–65

17. Tajiama Y, Rashed S, Okumoto Y et al (2007) Prediction of welding distortion and panel buckling of car carrier decks using database generated by FEA. Trans Join Weld Res Inst 36(1):65–71
18. Wang JC, Yin XQ, Murakawa H (2013) Experimental and computational analysis of residual buckling distortion of bead-on-plate welded joint. J Mater Process Technol 213(8):1447–1458
19. Wang JC, Murakawa H (2012) Fundamental study of buckling behavior in thin Plate butt welding by the inherent deformation method. In: Trends in welding research 2012: proceedings of the 9th international conference (ASM International), Chicago, pp 165–173
20. Deng D, Murakawa H, Liang W (2007) Numerical simulation of welding distortion in large structure. Comput Methods Appl Mech Eng 196(45/48):4613–4627

Chapter 3
Investigation on Thick-Plate Cutting of High-Strength Steel

Steel cutting plays a critical role in shipbuilding. Several thermal cutting techniques were developed. Most of these techniques, such as oxygen cutting, laser cutting, electron beam cutting, and plasma jet cutting, are based on moving the heat source along the desired trajectory over the sheet metal to achieve the objective of carving out a piece of metal to the desired shape and dimension.

Most studies on cutting temperature field adopted the conclusions published by various welding research studies. In fact, the oxygen cutting process is much more complex than welding or other cutting processes. First, it requires two heat sources and slag removal demands a substantial amount of energy. Because of the cutting slot, continuity of mass and heat is disrupted, which is not easy to model. To date, few analytical or numerical studies reported on composite heat sources [1].

In this chapter, we report an approach for developing a new model to describe composite heat sources based on the actual process of oxygen cutting. Numerical simulations of oxygen cutting were performed and corrected through experiments.

3.1 Research on Thermal Source Model of Oxygen and Acetylene Cutting

3.1.1 Thermal Source Model

Oxygen cutting comprises heating the surface to a prescribed temperature (combustion point) such that it can burn in oxygen via a gas flame (preheating flame). Then, cutting oxygen of high purity and high flow speed is filled into enable the iron in the steel to burn into ferric oxide slag in oxygen atmosphere. This process releases a lot of heat, through which the lower layer and the cutting edge of the steel can

© Science Press 2021

H. ZHOU and J. WANG, *FE Computation on Accuracy Fabrication of Ship and Offshore Structure Based on Processing Mechanics*, https://doi.org/10.1007/978-981-16-4087-2_3

be continuously heated to reach the combustion point to the bottom of the part. The cutting oxygen flows off the ferric oxide slag, forming slot to cut off the steel.

From the oxygen cutting process, we know that the flame first heats the steel to the combustion point, which is affected by moving a linear thermal source and a Gaussian thermal source.

3.1.2 Determination of Thermal-Flow Distribution Parameters

In oxygen cutting, the iron-oxygen combustion reaction provides a lot of energy, while blow-off melting slag demands a lot of energy. Therefore, the difference between both of these energies will be taken as the net energy, which is used in the calculation of the temperature field governed by Eq. (3.1). The energy will be assumed to be uniformly distributed along the thickness direction.

$$
\begin{aligned}
P_r' = P_r + P_h + P_f &= \eta_{gen} v w h \rho A_{gen} - v w h \rho c (T_{m1} - T_0) - v w h \rho L_f \\
&= v w h \rho \left[\eta_{gen} A_{gen} - c (T_{m1} - T_0) - L_f \right]
\end{aligned}
\tag{3.1}
$$

where P_r' is the net energy; P_r is the energy released in the combustion reaction; P_h is the energy necessary to heat the part to its melting point; P_f is the energy necessary to melt the part; η_{gen} is the iron combustion rate; A_{gen} is the iron-oxygen reaction heat; v is the cutting speed; w is the cutting width; h is the plate thickness; ρ is the material density; c is the plate specific heat capacity; T_0 is the primary temperature of the plate; L_f is the potential heat for material melting; and T_{m1} is the melting point temperature of the ferric oxide.

From the oxygen cutting process, we know that the flame first heats the steel to the combustion point, which is equivalent to acting on a semi-infinite body Gaussian thermal source. When the temperature of the steel plate reaches the combustion point, the heat released by the steel plate combustion will heat the lower layer plate, which is equivalent to acting on a semi-infinite body point thermal source. When the steel plate forms a cutting slot, the flame moves uniformly to cut off the steel plate gradually, which is equivalent to the common.

$$
T(\xi, y, z) = T_0 + \frac{Q}{2k\pi} \exp\left(\frac{-v\xi}{2\lambda}\right) \left[\sum_{-\infty}^{+\infty} (1/R_n) \right] \exp\left(\frac{-vR_n}{2\lambda}\right)
\tag{3.2}
$$

According to Eq. (3.2), the temperature field formed under the iron-oxygen combustion reaction effect can be expressed as follows:

$$
T - T_0 = \frac{P_r'}{2\pi k h} \exp\left(-\frac{vx}{2\alpha}\right) K_0 \left[r \sqrt{\frac{v^2}{4\alpha^2} + \frac{b}{a}} \right]
$$

$$= \frac{vw\rho}{2\pi k}[\eta_{gen}A_{gen} - c(T_{m1} - T_0) - L_f]\exp(-\frac{vx}{2\alpha})K_0\left[\sqrt{\frac{v^2}{4\alpha^2} + \frac{b}{a}}\right]$$

(3.3)

3.1.3 Quasi-stable Temperature Field Under the Effect of Thermal Sources

According to the superposition principle of temperature fields, the temperature field when cutting a thick plate with oxygen should be the sum of the temperature field under the effect of a Gaussian distribution thermal source and the effect of heat release from the iron-oxygen combustion reaction. From Eqs. (3.1) and (3.3), we can derive the following expression:

$$T_t = T_1 + T_r = \frac{2Q}{vc\rho}\frac{\exp\left(-\frac{z^2}{4\alpha t}\right)}{4\pi\alpha t^{1/2}}\frac{\exp[-\frac{r^2}{4\alpha(t+t_0)}]}{[4\pi\alpha(t + t_0)]^{1/2}}$$
$$+ \frac{vw\rho}{2\pi k}[\eta_{gen}A_{gen} - c(T_{m1} - T_0) - L_f]\exp\left(-\frac{vx}{2\alpha}\right)K_0\left(r\sqrt{\frac{v^2}{4\alpha^2} + \frac{b}{a}}\right) \quad (3.4)$$

The temperature field under flame is derived from a semi-large wireless body, while the temperature field under the effect of combustion heat release is obtained by assuming an infinite plate. In an actual thermal cutting process, it is necessary to correct Eq. (3.4). The thick and large part correction coefficient m_1 and thin-plate thickness correction coefficient m_2 are respectively introduced, so the quasi-stationary temperature field under the effect of the final double thermal source can be expressed as follows:

$$T_t = T_1 + T_r = m_1\frac{2Q}{vc\rho}\frac{\exp\left(-\frac{z^2}{4\alpha t}\right)}{4\pi\alpha t^{1/2}}\frac{\exp[-\frac{r^2}{4\alpha(t+t_0)}]}{[4\pi\alpha(t + t_0)]^{1/2}}$$
$$+ m_2\frac{vw\rho}{2\pi k}[\eta_{gen}A_{gen} - c(T_{m1} - T_0) - L_f]\exp\left(-\frac{vx}{2\alpha}\right)K_0\left(r\sqrt{\frac{v^2}{4\alpha^2} + \frac{b}{a}}\right)$$

(3.5)

3.1.4 Determination of Parameters in Thermal Source Model

The parameters in the thermal source model are mainly determined by the cutting process parameters, which mainly include the torch type, nozzle number, oxygen pressure, gas cutting speed, application of preheating flame, inclined angle between the nozzle and part, and the distance from the nozzle to the part surface. In a specific numerical simulation process, the process parameters directly determine the thermal energy input Q and the energy absorption efficiency coefficient K in the thermal source model.

3.2 Simulation of Rack Oxygen-Acetylene Cutting

3.2.1 Material Parameters of Temperature Properties

The physical performance parameters of the metal materials such as specific heat capacity, conductivity coefficient, elastic constant, and yielding stress vary with temperature. In the cutting process, part of the structure is heated to an ultra-high temperature. The change of temperature of the entire welding part is extremely violent. Therefore, in the numerical calculation of the cutting temperature and stress field, the change rule of all physical performance parameters of the materials with temperature should be determined.

One of the current problems with the technology of finite element simulation is the insufficient amount of data on thermal-physical performance parameters of the materials. Because of the limited number of measurement methods, values of parameters related to the thermal-physical performance such as specific heat capacity, conductivity coefficient, and density for the materials at high temperature, especially at temperatures approaching the melting temperature and under melting conditions, are still missing, and non-linear calculation is not applicable. At present, some data at high temperature are normally acquired through tests and linear interpolation. The tests, e.g., high-temperature tension test, always involve a significant effort and require high-end equipment, which just acquires a reduced amount of data. Therefore, interpolation on this basis will produce incorrect results.

Thermal-physical performance parameters of NV E690 material were calculated by using simulation software, which covered a wide range of material's properties. Table 3.1 shows the data after calculation revision.

3.2.2 Finite Element Model of Rack

Figure 3.1 shows an offshore jack-up platform leg driven rack. Its dimension parameters are as follows: thickness, 127 mm; teeth distance, 250 mm; teeth top height,

Table 3.1 Thermal-physical performance parameters of NV E690

Temperature (°C)	20	100	500	800	1200	1500	2000
Conductivity coefficient/($W \cdot m^{-1} \cdot K^{-1}$)	50.17	47.36	36.32	29.96	31.77	37.60	41.14
Density/($kg \cdot m^{-3}$)	7.86	7.84	7.70	7.65	7.47	6.76	6.76
Elastic modulus/MPa	2.09	2.14	1.78	1.45	0.99	0	0
Cutting modulus/MPa	81.20	78.95	68.08	54.68	17.72	0	0
Poisson ratio	0.29	0.31	0.33	0.36	0.37	0.5	0.5
Thermal expansion coefficient/°C^{-1}	12.56	13.09	15.72	20.72	24.42	24.42	24.42
Mole volume/mol	7.09	7.11	7.23	7.30	7.45	8.23	8.23
Linear expansion coefficient/K^{-1}	0.006	0.096	0.66	0.97	1.70	1.70	1.70
Volume modulus/Pa	165.56	164.07	155.28	142.03	120.82	118.45	118.45
Enthalpy/J	763.69	1176.71	14277.8	29216	44164	59194	59194
Specific heat capacity/($J \cdot kg^{-1} \cdot K$)	25.58	27.15	39.15	58.18	37.58	91.58	91.58

Fig. 3.1 Geometrical model
of rack

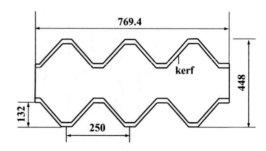

Fig. 3.2 Finite element
model of rack

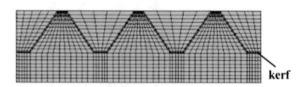

75 mm; total teeth height, 132 mm; teeth angle, 30 degrees; and rack width, 448 mm.
The length of the entire rack is 103667 mm. The entire rack was welded from several
sections of the same structure with a length of 769.4 mm. Each section includes 3
gear teeth.

To simplify the calculation model, we selected a section of three 769.4-mm-long
teeth as the research object. We simulated the condition by simultaneously cutting
both edges of the rack. To this end, it was necessary to ensure that the load and
structure were symmetrical. Then, we took 1/2 of the structure to establish the model
and avoid that the results were not collected in the strain analysis of the structural
stress. The process of establishing the rack model adopted a method from plane to
solid. The total quantity of units of the finite element model was 22590, and the total
quantity of nodes was 26160, as shown in Fig. 3.2.

In the case of cutting stress and strain analysis of the rack by indirect coupling
method, analysis on the temperature field of the rack should be first carried out. After
establishing the finite element model and input physical performance parameters of
the material, and finishing the boundary condition and loading, the analysis should
be started.

Figure 3.3 shows the temperature distribution contours of the cutting temperature
field.

Figure 3.3a shows the temperature field at the beginning of the cutting. The highest
temperature occurred on the surface of the cut part. The fact that the temperature
of the bottom part reached 1730 °C indicated that the temperature at the beginning
of the cutting had already exceeded the material melting point, which guaranteed
normal primary cutting. Figure 3.3b depicts the temperature field at 100 s. Note from
the temperature contours that the highest temperature was 2048 °C at the point at
which the thermal source was met. This is because, before the thermal source was
met, the material at the cutting seam had already burned in the former sub-step for
the following cutting seam area. This was equivalent to a "preheating" process, in

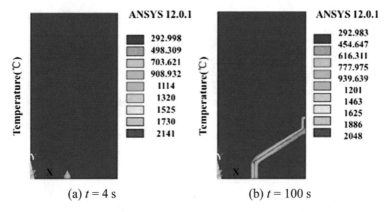

Fig. 3.3 Distribution of temperature fields at different cutting time instants

which the highest temperature is lower than the highest temperature in the primary cutting. In the primary cutting, the cutting speed was slightly slowed down to ensure normal cutting and guaranteed the head cutting of the primary cutting. Therefore, the heating time at the primary cutting seam was longer, and the temperature was the highest in the following cutting process.

Figure 3.4 shows that at a certain time instant after cutting, the cutting thermal source cut stably and the thermal influence area was also stable. This indicates that the cutting part had already entered quasi-stationary condition. Figure 3.4b shows the distribution of temperature field after the cutting process was completed. From the temperature contours of the cutting part at different time instants, it can be concluded that the thermal influence area of the material was small. Two reasons can be attributed to this phenomenon: the thermal conductivity of the material and the shape of the cut part.

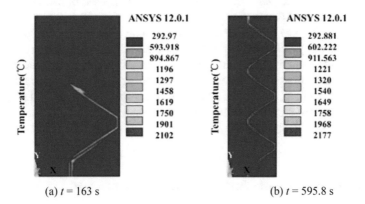

Fig. 3.4 Distribution of temperature fields at different cutting time instants

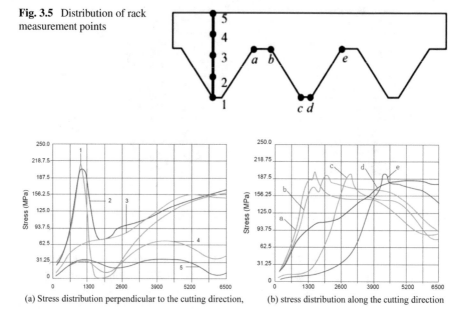

Fig. 3.5 Distribution of rack measurement points

(a) Stress distribution perpendicular to the cutting direction, (b) stress distribution along the cutting direction

Fig. 3.6 Time history of rack cutting stress contours

3.2.3 Calculation of Stress and Strain in Thermal Rack Cutting

Several typical points were selected to investigate the changing conditions of the stress on the rack over time. The distribution is shown in Fig. 3.5.

Figure 3.5 shows five points respectively selected on the cutting direction and the one perpendicular to it. The stress changing curve composed of the five points was then checked at three directions. The results are shown in Fig. 3.6.

Note from Fig. 3.6a that the residual stress mainly concentrated on the teeth top. At the part far away from the cutting seam, the residual stress decreased dramatically. Note also that the thermal influence area of NV E690 was small and the maximum value of the residual stress on the teeth top was 166 Mpa, which is significantly lower than the yield strength. As shown in Fig. 3.6b, the residual stress on the teeth top and teeth root along the cutting direction basically coincided, and the residual stress on the teeth top was slightly higher.

3.3 Optimization of Cutting Process Parameters

We aimed to optimize the three cutting parameters, namely total cutting thermal energy input, radius of the thermal source model, and cutting speed.

Table 3.2 Combinations of different values of cutting thermal energy input Q and radius of thermal source model r

Process parameter	Combination				
	1	2	3	4	5
Q	$0.8Q_0$	Q_0	$1.2Q_0$	$1.3Q_0$	$1.8Q_0$
r	$0.8r_0$	r_0	$1.2r_0$	$1.3r_0$	$1.25r_0$

Note $Q_0 = 2.8E + 3$ J, $r_0 = 35$ mm, $v_0 = 2.4$ mm/s

Table 3.2 shows the combinations of the five groups of cutting thermal energy input Q and radius of thermal source model r.

After investigating the temperature of the cutting seam surface of the first tooth after cutting, we obtained the results presented in Fig. 3.7. Note from this figure that when increasing the input energy, the temperature of the cutting seam surface also increased. In addition, while the input energy continued to increase together with the radius, the surface temperature of the cutting seam under the 4-th combination decreased. This indicates that in the cutting process, when the proportion of the input energy and radius of thermal source increases, the influence of the radius of the thermal source model on the highest temperature of the cutting seam is greater than that of the input thermal energy. By combining the five cases, we can identify when the input thermal energy is larger than a certain value. The influence of the input thermal energy on the highest temperature of the cutting seam constitutes the leading effect on that. By comparing combinations 2 and 3, we can determine when the total thermal energy input would be higher than $1.2Q_0$. At this point, the highest temperature of the cutting seam surface would reach 2500 °C. At such temperature, not only the metal at the cutting seam would be cut off, but also the metal near the cutting seam would be melted, resulting in the collapse of the teeth edge finally cut. A comparison of combination cases 2 and 4 shows that although the temperature for both cases basically coincides, combination 4 is $0.3Q_0$ higher than combination 2, which implies an extreme waste of cutting gas.

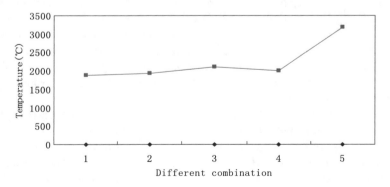

Fig. 3.7 Highest surface temperature for cases with different Q, r

Table 3.3 Combinations of different values of cutting thermal energy input and cutting speed

Process parameter	Combination				
	1	2	3	4	5
Q	$0.8Q_0$	Q_0	$1.2Q_0$	$1.3Q_0$	$1.8Q_0$
v	$0.8v_0$	v_0	$1.2v_0$	$1.3v_0$	$1.8v_0$

Note $Q_0 = 2.8\text{E} + 3$ J, $r_0 = 35$ mm, $v_0 = 2.4$ mm/s

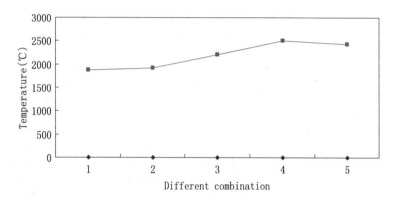

Fig. 3.8 Highest surface temperature diagram for different values of Q, V

Table 3.3 shows the combinations of five groups of cutting thermal energy input and cutting speed.

After investigating the highest temperature of the cutting seam surface after cutting the first teeth, we obtained the results presented in Fig. 3.8.

As shown in Fig. 3.8 for the first four groups of data, increasing the proportion of input energy and cutting speed increased the highest temperature of the cutting seam surface. Concerning the fifth group, although the input thermal energy was greatly increased as well, when the speed was increased too much, the cutting seam could not be heated completely within the resulting short period of time so the trend of temperature gain decreased. Note also from Fig. 3.8 that the input thermal energy was $1.2Q_0$. Although the cutting speed correspondingly increased to $1.2v$, the highest temperature of the cutting seam surface had already reached 2200 °C, and it may have also caused the collapse of the teeth edge in the cutting process.

Table 3.4 provides data of the combinations of the five groups of radius of the thermal source model and cutting speed.

After investigating the highest temperature of the cutting seam surface of the first teeth after cutting, we obtained the results presented in Fig. 3.9.

Concerning the first four groups of data in Fig. 3.9, we can see that for certain values of cutting thermal energy input, the highest temperature of the seam surface generally increased. As the radius of the thermal source decreased, although the cutting speed increased, the highest temperature of the cutting seam surface still increased. This indicates that the influence of the radius of the thermal source model

Table 3.4 Combinations of different values of cutting thermal energy input and radius of thermal source model

Process parameter	Combination						
	1	2	3	4	5	6	7
r	$1.3r_0$	$1.2r_0$	$1.1r_0$	r_0	$0.9r_0$	$0.8r_0$	$0.7r_0$
v	$0.7v_0$	$0.8v_0$	$0.9v_0$	v_0	$1.2v_0$	$1.4v_0$	$1.6v_0$

Note $Q_0 = 2.8\text{E} + 3$ J, $r_0 = 35$ mm, $v_0 = 2.4$ mm/s

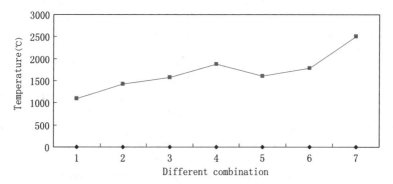

Fig. 3.9 Highest surface temperature of diagram for different values of r, v

on the highest temperature of the cutting seam surface was greater than that of the cutting speed. However, when the cutting speed increased to $1.2v_0$, the highest temperature of the cutting seam surface slightly decreased. This is a major advantage concerning the cutting speed in this combination. Then, with the continuous decrease of r, the cutting speed increased, and the radius r exhibited a major effect again, because the smaller the value of r was, the closer it was to the point thermal source, and the higher the highest temperature of the corresponding cutting seam surface.

3.4 Conclusion

(1) Based on the analysis of the cutting residual stress mechanism and cutting thermal source model, with the consideration of the process characteristics and processing parameters of the oxygen-acetylene flame cutting, we established a complex thermal source model of pre-heating thermal source and combustion reaction thermal source superposition. The model was based on the "birth-and-death" element technology, introducing the thickness correction coefficient, which was suitable for high-strength and large-thickness rack cutting. We also presented the thermal input distribution model of a new complex thermal source and determined the thermal source shape parameters by optimized design method.

(2)   On the basis of calculation and correction of the high-temperature physical performance of high-strength NVE690 steel by JMATPRO software, simulation calculations were carried out on the cutting of a jack-up platform leg rack. The results show that the thermal conductivity rate of NV E690 was lower and that, during the cutting and cooling processes, the thermal conductivity coefficient (such as ventilation) should be enlarged and the rack cooling speed should be sped up. After cutting, the residual stress should mainly concentrate on the teeth root and teeth top, where the residual stress can reach its maximum. Setting the cutting input energy Q to $2.8E + 3$ J, the cutting speed V to 2.0 mm/s, and the radius of the thermal source r to 35 mm, normal cutting can be guaranteed without non-through cutting of the rack or collapse of the edge.

(3)   Research on the optimization of the stress and strain was conducted for variables such as cutting speed, for the thermal source model, and for other related parameters. Finally, the proper cutting speed and related thermal source parameters were proposed in the cutting process, and a certain theoretical instruction was provided for the site cutting. The research results shows that when v is fixed, the highest temperature of the cutting seam surface increases with increasing Q. When the input thermal energy Q is larger than $1.3Q_0$, edge collapse may be easily caused; when r is fixed, and the cutting speed is smaller than $1.3v_0$, the speed change does not have significant influence on the highest temperature of the cutting seam surface, and the main factor is still the input thermal energy Q; when the cutting speed is larger than $1.3v_0$, the cutting speed has an extremely great influence on the highest temperature of the cutting seam surface; when Q is fixed, the influence of r and v on the highest temperature of the cutting seam surface presents an alternative leading effect.

Reference

1. Zhou Bo, Liu Yu-jun, Tan Soon-Keat (2013) Efficient simulation of oxygen cutting using a composite heat source model. Int J Heat Mass Transf 57:304–311

Chapter 4
Hull Plate Bending with Induction Heating

To ensure excellent hydrodynamic performance, ship structures are always designed and fabricated with a curved hull, which is obtained by plate formation techniques applied in shipyards. In particular, bulbous bow and stern sections are assembled by thick plates with complex double curvature, such as concave or convex profile. Thus, the plate formation procedure will significantly influence the dimensional precision and external appearance of ship structures, in addition to fabrication cost and schedule. In general, plate formation or plate bending during ship hull manufacturing can be achieved by means of a cold formation process, such as pressing and rolling, and a hot formation process with oxygen-acetylene flame heating, laser heating, and induction heating.

Induction heating can be considered as a physical process of heating on magnetic materials, such as iron and steel, with high permeability owing to electromagnetic induction [1]. In particular, induction heat is produced inside a specimen by the eddy currents, which is the result of a high-frequency oscillatory current of induction coil and a rapidly alternating magnetic field. Different out-of-plane bending deformations of different materials and thickness plates were obtained by multiple line heating with high-frequency induction heating equipment. In this chapter, the experiments, measurements, and numerical simulations conducted on a saddle-shape plate are introduced.

4.1 Experimental Procedure and Measurement

Generally, an induction heating system consists of an electromagnet, such as a ship plate steel, an induction coil, a water cooling recycle unit, and an electronic oscillator that passes a high-frequency alternating current (AC) through the electromagnet. In the basic setup, AC power supply will provide electricity with low voltage but very large current and high frequency. The induction coil is usually made of copper tubing

© Science Press 2021
H. ZHOU and J. WANG, *FE Computation on Accuracy Fabrication of Ship and Offshore Structure Based on Processing Mechanics*,
https://doi.org/10.1007/978-981-16-4087-2_4

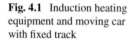

Fig. 4.1 Induction heating
equipment and moving car
with fixed track

and cooled fluid; its diameter, shape, and number of turns all influence the efficiency
and field pattern. In this study, the advanced induction heating equipment shown in
Fig. 4.1 with high frequency (30–100 kHz) was employed. Its working power, output
current, and output voltage are 25 kW, 1060 A and 24 V, respectively. An additional
water cooling system with 12 kW of power, 50,000 kcal/h of cooling capacity, and
12.5 m^3/h of circulating water quantity was also applied. To ensure steady heating
speed during the plate-bending experimental procedure, a moving car with fixed
track was used.

The dominant physical behavior, transient temperature, and thermal cycle with
moving induction coil were measured with a K-type thermal couple in real time.
We aimed at examining the relation between induction heating velocity and heated
maximal temperature. After cooling down to ambient temperature, a three-coordinate
measuring machine was employed to obtain the geometrical profile and out-of-plane
bending deformation.

4.1.1   Temperature Measurement During Induction Heating

After induction heating on plate, the bending moment, which was generated by non-
uniform compressive plastic strain in transverse direction through the thickness direction, may produce out-of-plane bending deformation. The underlying mechanism of
this mechanical response was temperature gradient through the plate thickness direction. No out-of-plane bending deformation occurred for an identical distribution of
temperature through the plate thickness direction.

Measurements and data analysis of heating temperatures were first carried out
during an experiment of plate bending with induction heating. Specifically, a K-type

thermal couple with platinum–rhodium and high-precision transient-temperature measurement device was employed to obtain the thermal cycles of several measured points. A thermal couple employed as a temperature sensor consists of two dissimilar electrical conductors with a formation of electrical junction. It can generate a temperature-dependent voltage as a result of thermal-electrical effect or thermal electromotance.

To examine the thermal behavior during induction heating, we used a test steel plate with 8 thermal couples, as shown in Fig. 4.2. During the experiments, the transient temperature measuring device can simultaneously obtain and save temperature data of 8 connected thermal-couples at 16 Hz/s of measuring frequency. The locations of these 8 thermal couples were sequentially designed along the direction of the moving induction coil. The K-type thermal couple with platinum–rhodium was fixed to specific locations of the plate by a special tack welding machine; the locations are indicated in Fig. 4.2. The other side was connected to the temperature measuring device.

With the moving car and its fixed track, 8 cases with different moving speeds of the induction heating coil, as summarized in Table 4.1, were carried out. Figure 4.3 shows the transient temperature distribution when the induction coil with a moving

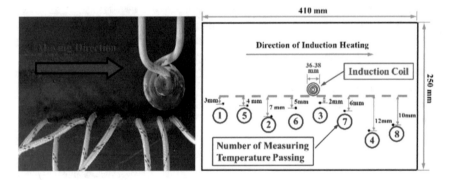

Fig. 4.2 Locations of points measured with thermal couples

Table 4.1 Parameters of induction heating

Case No	Moving distance (mm)	Time (s)	Moving speed (mm/s)	Linear energy (kJ/mm)
1	250	102	2.451	2.14924
2	250	269	0.929	6.28141
3	110	99.5	1.106	5.09996
4	80	90	0.889	12.40079
5	220	243	0.905	11.02293
6	102	166	0.614	13.45533
7	220	369	0.596	23.99232
8	220	370	0.595	27.23312

Fig. 4.3 Transient temperature field with moving induction coil

speed of 0.905 mm/s (case 4) was heating the examined ship plate and the thermal couples were passed during experiment.

Given that the power of induction heating was fixed to 25 kW, different energy values per length could be calculated. For case 4 (moving speed of the induction heating coil of 0.0905 mm/s), Fig. 4.4 shows the thermal cycles of measured points away from the induction heating line, as shown in Fig. 4.2. Thermal couples at points 1, 4, and 7 had connection problems, providing no temperature signals. The relation between linear energy values of induction heating as summarized in Table 4.1 and measured data of heated maximal temperatures can be established, as shown in Fig. 4.5. Vertical direction means linear energy, which is determined by the power input and moving speed of the induction heating coil, as summarized in Table 4.1; upright direction means measured maximal temperature for different thermal couples with different linear energy values and moving speeds of induction heating. It can be concluded that the maximal temperature of measured point is decreasing while linear energy is decreasing or the measured point is moving further away from the heating line.

Fig. 4.4 Measured thermal cycles of points away from the heating line

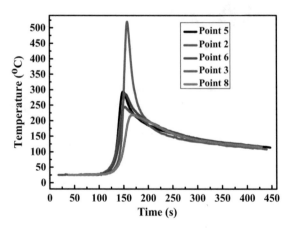

Fig. 4.5 Relation between measured maximal temperature of thermal couple and linear energy of induction heating

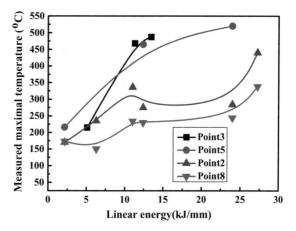

4.1.2 Measurement of Bending Deformation

After cooling down, the heated ship plate presented out-of-plane bending deformation due to the mechanical response caused by heating. The three-coordinate measuring machine shown in Fig. 4.6 was employed to obtain the coordinates of the measured point cloud, which was then used to establish the geometrical profile through measured data analysis.

Fig. 4.6 High-precision three-coordinate measuring machine (2 um)

In particular, the ship plate with 10-mm thickness after induction heating shown in Fig. 4.7 was examined. After surface cleaning to avoid measurement disturbance, coated abrasive for rust removal was applied on the measurement surface. Then, the measured ship plate was fixed to the measuring platform by fixture or clamping of the three-coordinate measuring machine. This ensured the required measurement precision and completion of the measuring procedure, as shown in Fig. 4.8.

Figure 4.8 also shows the measurement procedure with computer control, in which the referenced coordinate system had be confirmed before measuring. The number and locations of measured points can be fixed according to the balance of measurement cost and required measurement precision. Thus, the distance between each adjacent point is eventually 15 mm with about 2-um tolerance.

Data analysis enables visualization of out-of-plane bending deformation through post-processing, as shown in Fig. 4.9. Note from the measured contour that the deformed ship plate presented an evident angular distortion in the direction perpendicular to the induction heating line. Figure 4.10 depicts the out-of-plane bending deformations of points on the line perpendicular to the induction heating line, which was located very close to the middle region of the examined ship plate.

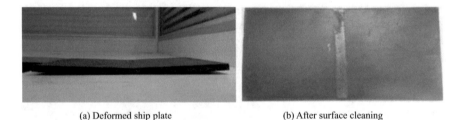

(a) Deformed ship plate   (b) After surface cleaning

Fig. 4.7 Examined ship plate with out-of-plane bending deformation

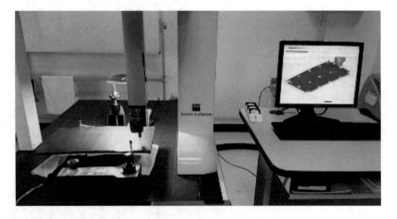

Fig. 4.8 Procedure of coordinate measuring with computer control

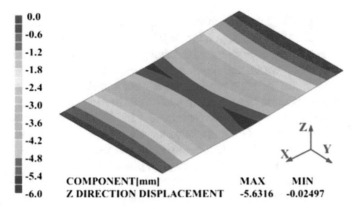

COMPONENT[mm]   MAX   MIN
Z DIRECTION DISPLACEMENT   -5.6316   -0.02497

Fig. 4.9 Measured contour of out-of-plane bending deformation (10-mm thickness)

Fig. 4.10 Out-of-plane
bending deformation due to
linear induction heating

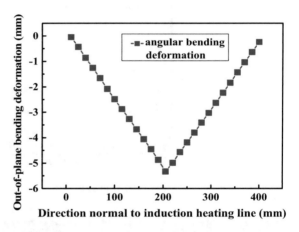

4.2 Measurement and Computational Analysis of Saddle Plate

Owing to the cost and time consumption, computational analysis with advanced computer and numerical technology was accepted and employed for out-of-plane welding deformation prediction during plate bending with induction heating. In general, there are two elementary approaches: thermal elastic–plastic FE computation to represent the entire thermal transfer behavior and its generated mechanical response, and elastic FE computation focused on residual bending deformation with bending moment caused by induction heating.

In previous studies on single linear induction heating on plate bending, the underlying physical phenomenon was observed and transverse bending deformation was measured with a three-coordinate measuring machine. In this section, an object plate with double curvature for shipbuilding was examined. Owing to the designed geometry of ship hull plate, two typical bending plates with saddle shape (opposite curvatures) were achieved through multiple linear induction heating.

As a practice for plate bending with saddle shape, a ship plate with 300 mm in length, 200 mm in width, and 6 mm in thickness was examined. Its yield stress was about 235 MPa. The plate bending experiment was then carried out with the aforementioned induction heating equipment with a moving speed of the induction coil of approximately 13.75 mm/sec for lines 1–4 and 23.58 mm/sec for lines 5 and 6. The sequence and route of these six induction heating lines are shown in Fig. 4.11.

After cooling down, the bending deformation was measured with an advanced three-coordinate measuring machine with high precision (global classis SR series with 2-um tolerance). Plate surface clearance was first carried out to ensure the measured accuracy. Then, the plate was fixed on the platform, as shown in Fig. 4.12. Three points were next selected and marked to create a referenced plane. The new

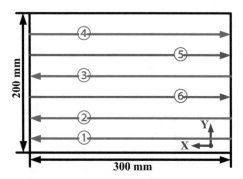

Fig. 4.11 Line heating pattern on Q235 plate (6 mm)

Fig. 4.12 Deformed plate with saddle shape on measurement platform with fixture

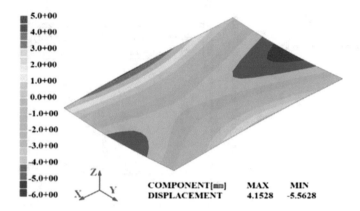

Fig. 4.13 Contour and distribution of measured out-of-plane displacement (saddle shape)

coordinate of the bending plate in the thickness direction was measured and the deformed shape of the examined plate was established through processing and graphical visualization, as shown in Fig. 4.13. In particular, Fig. 4.13 shows the contour and distribution of the measured out-of-plane bending displacement. Note that the examined plate bended upward in the heating line direction, while it bended downward in the direction perpendicular to the heating line. In other words, the bending curvatures in length and width directions exhibited opposite sign. This is the dominant feature of a saddle-type bending plate.

To examine the out-of-plane deformation characteristic of the bending plate with saddle shape in detail, lines 1 and 2, shown in Fig. 4.14, were selected and out-of-plane displacements of points on lines 1 and 2 were then plotted, as shown in Fig. 4.15. It can be clearly concluded that the bending trend of points on line 1 (longitudinal direction) is overall upward and the bending trend of points on line 2 (transverse direction) is conversely downward.

Fig. 4.14 Positions for out-of-plane bending distortion measurement with multiple linear induced heating

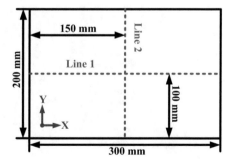

Fig. 4.15 Measured
out-of-plane welding
distortions of points on lines
1 and 2

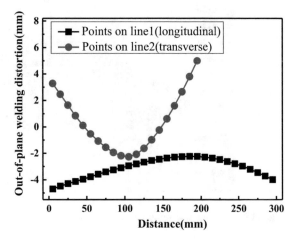

4.2.1   Thermal Elastic–Plastic FE Computation

A solid brick element model shown in Fig. 4.16 was made for thermal elastic–
plastic FE computation. The geometrical profile was identical to the experiment
specimen. The total number of nodes and elements were 32,513 and 28,800, respectively. Non-linear temperature-dependent material properties shown in Fig. 4.17 were
employed. During the thermal analysis, a body heat source with uniform flux density
was employed to model the induction heating. Complex electric–magnetic physical
behavior was avoided. Considering the power and moving speed of the induction
heating in the previous experiment, the heat input per length was obtained and applied
to the heated region, which was determined by the size of the induction heating coil
and penetration depth. The rigid body motion shown in Fig. 4.16 was fixed as the

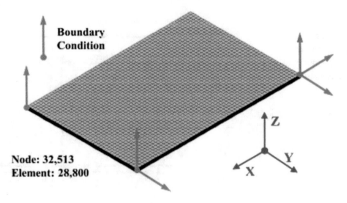

Fig. 4.16 Solid brick elements of the examined bending plate

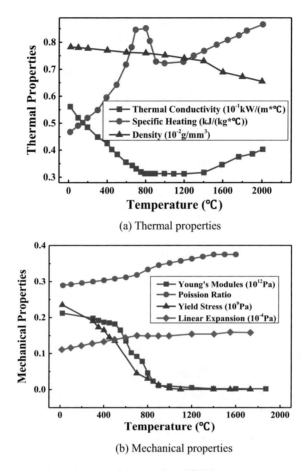

(a) Thermal properties

(b) Mechanical properties

Fig. 4.17 Temperature-dependent material properties of Q235

boundary condition to predict the out-of-plane bending deformation according to the actual experiment.

An in-house code was employed for thermal elastic–plastic FE computation. Figure 4.18 shows the transient temperature distribution during the induction heating. Note that the proposed body heat source can represent a similar temperature field to that of the experiment. In particular, the region, whose highest transient temperature was approximately 700 °C, exhibited a very similar size to that of the induction heating coil. The aforementioned highest temperature approximates the Curie point without exceeding it.

After cooling down, the residual plastic strain and out-of-plane bending deformation were computed, as shown in Fig. 4.19, where the bending plate with convex curvature in the length direction and concave curvature in the width direction are represented. This bending type was generated owing to the compressive shrinkage force in the induction heating direction. The examined plate lost the stability with

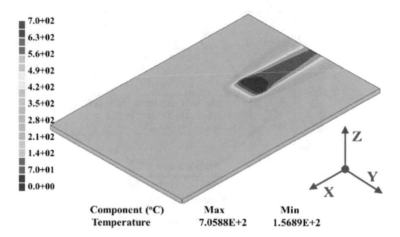

Fig. 4.18 Transient temperature distribution during induction heating

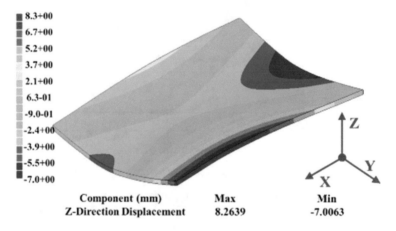

Fig. 4.19 Contour plot of out-of-plane bending deformation after cooling down

buckling behavior in saddle mode. A comparison of the magnitude of the out-of-plane bending deformation with the corresponding measurements will be presented later on.

4.2.2 Evaluation of Bending Moment

Although the thermal elastic–plastic FE computation can represent the buckling behavior of plate under induction heating and predict the magnitude of out-of-plane bending deformation, it requires a great deal of computer resources and computing time. Elastic FE computation can be employed to predict the buckling behavior more effectively and the magnitude of out-of-plane bending deformation provided that the shrinkage force and bending moment caused by induction heating can be previously evaluated [2].

Figure 4.20 shows the contour distribution of plastic strain along the induction heating direction. The shrinkage force can be evaluated through integration by Eq. (4.1). The distribution of plastic strain perpendicular to the induction heating direction was also obtained, as shown in Fig. 4.20. Given that the plastic strain presented a gradient in the thickness direction, the bending moment was generated and its magnitude could be evaluated through Eq. (4.2) the shrinkage force and bending moment caused by induction heating with different moving speeds (13.75 mm/s for lines 1–4 and 23.58 mm/s for lines 5 and 6) were evaluated through

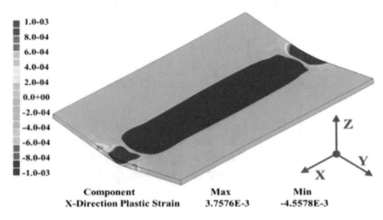

(a) Contour distribution of plastic strain in X direction

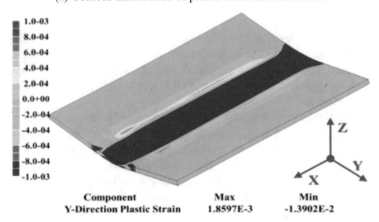

(b) Contour distribution of plastic strain in Y direction

Fig. 4.20 Contour distribution of plastic strain caused by induction heating

Table 4.2 Shrinkage force and bending moment caused by induction heating

Moving speed (mm/s)	Longitudinal shrinkage force (N)	Transverse bending moment (N·m)
13.75 (lines 1–4)	−97,221.6	134,114.4
23.58 (lines 5 and 6)	−94,663.8	173,464.2

thermal elastic–plastic FE computation and integration of computed plastic strain. The results are summarized in Table 4.2. These results were applied for subsequent elastic FE analysis as input load.

$$F_{shrinkage} = \iint E \times \varepsilon_x^{plastic} dy dz \tag{4.1}$$

$$dM = \left(z - \frac{h}{2}\right) \times dF = \left(z - \frac{h}{2}\right) \times E \times \varepsilon_y^{plastic} dA$$

$$M_{bending} = \iint \left(z - \frac{h}{2}\right) \times E \times \varepsilon_y^{plastic} dy dz \tag{4.2}$$

where, $\varepsilon_x^{plastic}$ and $\varepsilon_y^{plastic}$ are the plastic strains generated by induction heating in the longitudinal and the transverse directions, E is the young's modules, h is the thickness of the plate with induction heating, and x, y, z are the induction heating direction, transverse direction and thickness direction, respectively.

4.2.3   Elastic FE Computation

When the shrinkage force and bending moment caused by induction heating were obtained and applied as the mechanical response, elastic FE analysis based on shell-element model, shown in Fig. 4.21, was carried out. Different colors were used to distinguish the parts. An induction heating line was located between adjacent parts with different colors.

The shell elements allowed for the examined plate to be meshed with coarse elements regardless of the induction heating region. In addition, less computer memory and computing time were required. The total number of nodes and elements were 2,501 and 2,400, respectively. The line between adjacent parts with different colors was heated by induction heating according to the prescribed experiment. The rigid body motion, also shown in Fig. 4.21, was constrained to be boundary condition during mechanical analysis.

Figure 4.22 shows the contour plot of out-of-plane bending deformation after completion of induction heating. The straight line indicates the position of the induction line heating, which was identical to the experimental observation. A saddle-type buckling bending plate was represented. The induction heating direction exhibited a convex bending curvature and the direction perpendicular to the induction heating

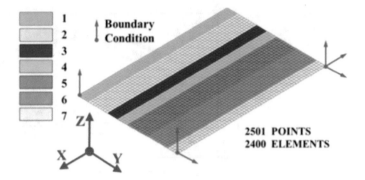

Fig. 4.21 Shell elements of the examined bending plate

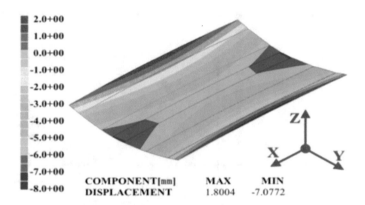

Fig. 4.22 Contour plot of out-of-plane bending deformation with elastic FE analysis

presented a concave bending curvature [3]. It can be concluded that the examined plate with 6 mm in thickness was buckled owing to the compressive shrinkage force in the length direction caused by induction heating. The bending moment is assumed to play the role of external disturbance that triggered the occurrence of buckling behavior when the compressive shrinkage force was greater than the critical buckling load [4].

For further investigation, out-of-plane bending deformations of points on lines 1 and 2, shown in Fig. 4.14, were summarized and compared in Fig. 4.23. The plots in this figure were obtained through measurement, thermal elastic–plastic FE analysis, and elastic FE analysis. Figure 4.23 shows that points on line 2 have a concave bending profile and their magnitudes have good agreement with each other. Not only the bending shape but also the magnitude of the considered points resulting from thermal elastic–plastic FE and elastic FE computations had consistent features with measurements. However, elastic FE computation has many more advantages as an effective and practical computational approach. It requires less computer resources and computing time compared to thermal elastic–plastic FE computation, as shown in Table 4.3.

Fig. 4.23 Comparison of
out-of-plane bending
deformation

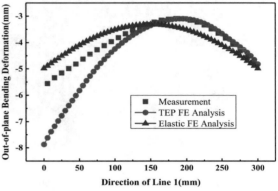

(a) out-of-plane bending deformations of points on line 1

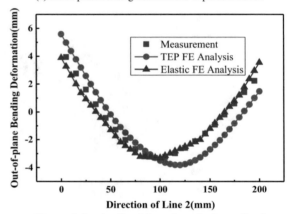

(b) out-of-plane bending deformations of points on line 2

Table 4.3 Comparison of
computational consumption

	Node number	Element number	Computing time (s)
TEP FE Analysis	32,513	28,800	3,007
Elastic FE Analysis	2,501	2,400	2.5

4.3 Measurement and Computational Analysis of Sail Plate

In shipyards, the common double curvature plate includes a sail-shape plate beside the
saddle-shape plate. Research on sail-shape plates by experimental and computational
methods is introduced next.

A plate with high tensile strength steel (AH36) with a yield stress of approximately
360 MPa was implemented to conduct plate bending deformation experiments with

the induction heating equipment above described. Its geometric dimensions were 300*200*8 mm (length*width*thickness). Figure 4.24 shows the detailed route and sequence of induction line heating. The distance between two lines is regular (50 mm in the length direction and 40 mm in the width direction). The numbers, arrow direction, and line types in the figure respectively indicate the heating sequence, coil moving direction, and different moving speeds (solid and dashed lines represent 19.1 and 21.1 mm/s respectively).

After cooling down, the test plate presented a sail shape, as shown in Fig. 4.25. Then, it was fixed on the platform by the aforementioned TMM to measure the out-of-plane bending deformation of the examined ship plate with a procedure similar to the one described in Sect. 4.1. Through post-processing of the measured data, the geometrical profile of the measured surface can be established and the out-of-plane displacement distribution can be visualized through different colors, as

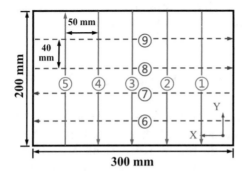

Fig. 4.24 Line heating route on an AH36 plate (8 mm)

Fig. 4.25 Sail-shape plate after cleaning up and fixing on measured platform

Fig. 4.26 Measured
out-of-plane displacement
distribution and contour of
sail-shape plate

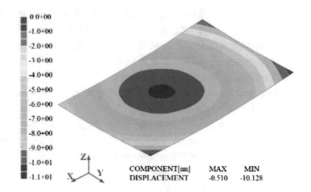

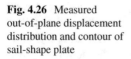

shown in Fig. 4.26. This figure also shows that the bending plate bended downward both in length and width directions, and the deformation reached its maximum, approximately 10 mm, at the middle of the plate.

4.3.1   TEP FE Analysis for Sail-Shape Plate

A solid-element model for TEP FE analysis is shown in Fig. 4.27. Taking into account the experimental equipment condition mentioned in Sect. 4.2, the in-plane element size was 5*5 mm, and the element length along thickness was 0.5 mm, as shown in Fig. 4.27. The model consisted of 42,517 nodes and 38,400 elements, and the boundary condition established 6 displacements at 3 nodes.

The plate bending deformation process was simulated through TEP FE analysis by considering large deformation and employing a heat input corresponding to induction in experimental heating condition as well as high-temperature thermal–mechanical

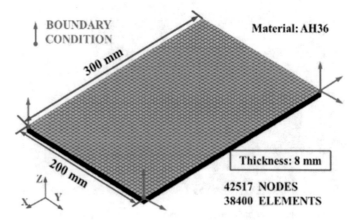

Fig. 4.27 Solid-element model for sail-shape plate

properties of material AH36. Figure 4.28 shows the computed transient temperature distribution when heating along the plate width direction. It also shows the computed maximum temperature distribution for the whole induction heating process; the route of the line heating can be clearly identified. The mechanical analysis results of the out-of-plane bending displacement with TEP FE analysis are depicted in Fig. 4.29. Note that a similar sail shape to that of the measurement results and identical curvature in length and width directions were obtained. The points on lines 1 and 2 shown in Fig. 4.14 were examined again. Their out-of-plane bending displacements are presented in Fig. 4.30. The calculation results of out-of-plane displacement are mostly consistent with the measured data in both length and width directions.

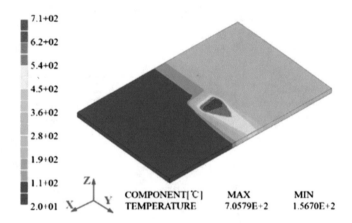

(a) Transient temperature distribution

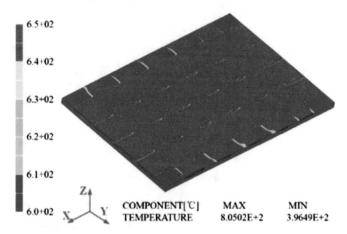

(b) Maximum temperature distribution for the whole process

Fig. 4.28 Computed temperature results for sail-shape plate from thermal analysis

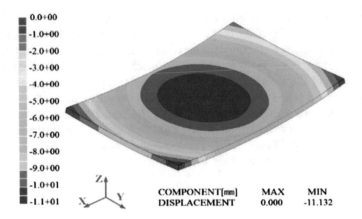

Fig. 4.29 Computed out-of-plane displacement contour with TEP FE analysis (sail-shape plate)

Fig. 4.30 Out-of-plane
displacement comparison
between measurement and
TEP FE analysis on two lines
(sail-shape plate)

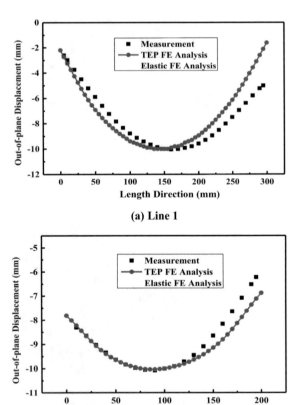

4.3.2 Elastic FE Analysis for Sail-Shape Plate

Owing to the different coil velocities in plate length and width directions, two groups of inherent deformation resulted from integrating the plastic strain results of TEP FE analysis, as shown in Table 4.4.

The model for elastic FE analysis (composed of 2501 nodes and 2400 elements) included shell elements, as shown in Fig. 4.31. The in-plane element size was 5*5 mm. The boundary conditions were similar to those of TEP FE analysis, with 6 DOFS fixed. The model of the plate was divided into 30 pieces with different colors because the bending plate with sail shape had 5 heating lines along the width direction and 4 heating lines along the length direction.

Figure 4.32 shows the distribution of computed out-of-plane displacement by elastic FE analysis. Note that the bending plate with sail shape can be clearly identified. Bending deformations on lines 1 and 2 (shown in Fig. 9) from both calculation methods were compared with the corresponding measurements, as shown in Fig. 4.33. Note that the computed results from elastic FE analysis agree well with the corresponding measurements and with the results from TEP FE analysis.

Table 4.4 Magnitudes of inherent deformation (material: Q235; thickness: 6 mm)

Velocities (mm/s)	19.1	21.1
Longitudinal shrinkage (mm)	−0.08402	−0.09269
Transverse shrinkage (mm)	−0.10890	−0.10608
Transverse bending (rad)	−0.03713	−0.03198
Longitudinal bending (rad)	−0.00472	−0.00505

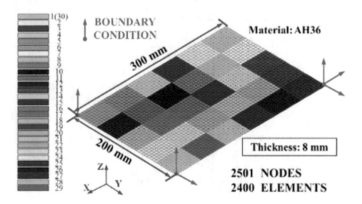

Fig. 4.31 Shell-element model for sail-shape plate

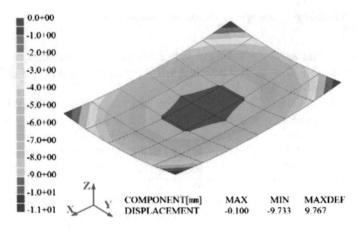

Fig. 4.32 Computed out-of-plane displacement contour with elastic FE analysis (sail-shape plate)

Fig. 4.33 Out-of-plane
displacement comparison
between measurements, TEP
FE analysis, and elastic FE
analysis on two lines
(sail-shape plate)

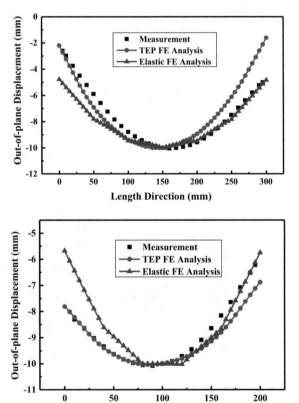

4.4 Discussions and Conclusions

According to our results on plate bending with induction heating and the comparison of out-of-plane displacement between calculated and measured results, TEP FE analysis and elastic FE analysis can be applied to predict plate bending deformation resulting from induction heating. However, the required computer resources, model complexity, and computing time of TEP FE analysis and elastic FE analysis are notably different. Relative data are given in Table 4.5. It can be concluded that the model for elastic FE analysis is simpler and consumes less computer resources and computing time.

Furthermore, although the calculated results are in good agreement with measurements, numerical simulation results cannot perfectly coincide with the measurement results. On the one hand, errors in experiments and corresponding measurements occur, e.g., the deviation of line heating position or the change of velocity during single line heating. On the other hand, numerical calculation implies simplification, e.g., when setting the boundary condition or the loading of inherent deformation. However, both calculation methods can predict the trend of out-of-plane plate deformation and the computed results are accurate enough with respect to the measurements. Concerning TEP FE analysis, adding the induction heating line constitutes a problem because re-modeling, re-selection of the heat source, and re-calculation of the temperature field and bending deformation are required. In this regard, elastic FE analysis is much more convenient owing to the constant inherent deformation. It only needs to add a process with loading inherent deformation. The accuracy of the elastic FE analysis results depends on the accuracy of the inherent deformation value, which can be obtained from the results of TEP FE analysis or from experiments. The calculation results are quickly obtained with elastic FE analysis based on inherent deformation theory. Therefore, the elastic FE analysis is more flexible and more suitable for the engineering problem of plate formation by induction heating.

A ship-plate formation experiment with high-frequency induction heating equipment was first carried out. The out-of-plane bending deformation was measured by an advanced TMM after cooling down the plate. Then, TEP FE analysis and elastic FE analysis were applied to simulate the induction heating process of plate formation and predict the out-of-plane bending deformation. The following conclusions can be drawn:

Table 4.5 Comparison of computation parameters for TEP analysis and elastic analysis

	Method	Nodes	Elements	Computing time (min)
Bending plate with saddle shape	TEP analysis	32,513	28,800	3007
	Elastic analysis	2501	2400	2.5
Bending plate with sail shape	TEP analysis	42,517	38,400	4140
	Elastic analysis	2501	2400	4

(1) High-frequency induction heating can be employed to obtain typical double-curvature hull bending plates of different thicknesses and materials through different line heating patterns.

(2) A three-coordinate measuring machine with high precision can accurately measure the magnitude and distribution of out-of-plane bending deformation. Thus, the geometrical profile of the measured surface is depicted by simple post-processing.

(3) TEP FE analysis and elastic FE analysis can be applied to predict the plate bending deformation due to induction heating. The calculation results agree well with the measurements.

(4) The model for elastic FE analysis is simpler, consumes less computer resources, and has higher efficiency. Furthermore, elastic FE analysis can quickly and accurately predict the bending deformation due to the constant inherent deformation and has engineering application value when adding the induction heating line.

References

1. Rudnev V, Loveless D, Cook RL, Black M (2017) Handbook of induction heating, 2nd edn. CRC Press.
2. Wang JC, Shibahara M, Zhang X, Murakawa H (2012) Investigation on twisting distortion of thin plate stiffened structure under welding. J Mater Process Technol 212(8):1705–1715
3. Wang J, Yin X, Murakawa H (2013) Experimental and computational analysis of residual buckling distortion of bead-on-plate welded joint. J Mater Process Technol 213(8):1447–1458
4. Wang JC, Rashed S, Murakawa H (2014) Mechanism investigation of welding induced buckling using inherent deformation method. Thin-Walled Struct 80:103–119

Chapter 5
Out-of-Plane Welding Distortion Prediction for Typical Welded Joints and Ship Structures

In contrast with other joining methods, welding is widely accepted for fabrication and assembly of metal structures owing to its practical and highly productive features [1]. In particular, fusion welding is a main welding method that is employed in the production of virtually all types of steel structures, such as marine structures, automobiles, trains, aircrafts, bridges, and pressured vessels.

The welding procedure plays an essential role during the fabrication of ships and offshore structures. It approximately consumes 50–60% of the fabrication schedule. Welding with complex processing parameters significantly influences the fabrication quality.

Because of extremely non-uniform heating, welding-induced distortion cannot be avoided during heating and cooling. It causes loss of dimensional control, structural integrity, difficult subsequent alignment with the adjacent component, and increase of fabrication costs owing to straightening [2]. Generally speaking, welding distortion can be divided into in-plane shrinkage and out-of-plane welding distortion. The latter is a combination of transverse bending (angular distortion), longitudinal bending, and welding-induced buckling, as shown in Fig. 5.1.

It is well known that transverse bending is the dominant phenomenon that generates out-of-plate welding distortion in case of thick-plate welded structures. However, with the use of relatively thin plates, not only conventional bending distortion but also welding buckling may become contributing factors to produce out-of-plane welding distortion, which is caused by different reasons [3]. Thus, clarifying the generation mechanism and mitigating out-of-plane welding distortion with an appropriate flame-heating process have become essential research issues.

© Science Press 2021

H. ZHOU and J. WANG, *FE Computation on Accuracy Fabrication of Ship and Offshore Structure Based on Processing Mechanics*, https://doi.org/10.1007/978-981-16-4087-2_5

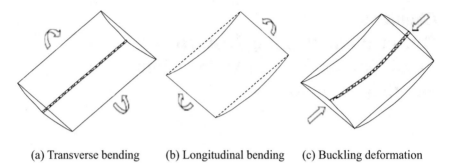

(a) Transverse bending   (b) Longitudinal bending   (c) Buckling deformation

Fig. 5.1 Typical out-of-plane welding distortions

5.1 Welding Distortion of Typical Fillet Welding

Among typical welded joints in ship and offshore structures, fillet welding is an essential joining procedure to assemble flange and web parts together. Thus, it is necessary to examine the mechanical response and out-of-plane welding distortion of fillet welded joints as a previous step.

5.1.1 Experimental Procedure and Measurement

A typical welded T-joint was assembled by the flange with 1200 mm in length, 200 mm in width, 9 mm in thickness. A web with 1200 mm in length, 100 mm in height, and 9 mm in thickness was selected. Its geometrical profile was 8 mm in leg length and approximately 2 mm in penetration, as illustrated in Fig. 5.2. Chemical elements contents of SS400 are listed in Table 5.1. GMAW with CO2 as shield gas was employed. The corresponding welding conditions are listed in Table 5.2.

After the welded joint cooled down to the ambient temperature, out-of-plane welding distortion was measured. The locations of the measured points are indicated in Fig. 5.3. Magnitudes of measured angular distortion were obtained and summarized in Table 5.3. Note that the angular distortion increased smoothly along the welding direction.

In this experimental study, two typical fillet welded joints with different geometrical profiles in cross-sections were considered first. Then, welded structures with two parallel stiffeners and two longitudinal and transverse stiffeners in a cross form were made. All employed materials for welding were SS400, whose chemical elements are given in Table 5.1.

Owing to its effective and practical features, CO_2 welding was applied for structure assembling. Welding induced out-of-plane distortion was measured after cooling down. Next, detailed dimensional information, welding process, and measurements of the aforementioned welded specimens are orderly presented and examined.

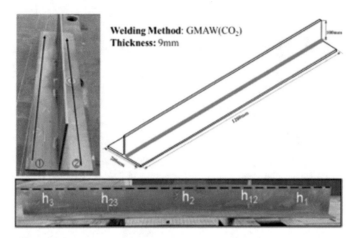

Welding Method: GMAW(CO_2)
Thickness: 9mm

h_3 h_{23} h_2 h_{12} h_1

Fig. 5.2 Specimen of experimental welded T-joint

Table 5.1 Primary chemical elements of SS 400 (mass percentage)

Chemical elements	C (max)	Si	Mn (max)	P (max)	S (max)
Percentage of mass (%)	0.17	Not controlled	1.40	0.045	0.045

Table 5.2 Welding condition for assembling of fillet welded joints

Welded joint	Current (A)	Voltage (V)	Heat input (J/mm)
T200 fillet welded joint	230	26-24	1495
T300 fillet welded joint	270	27-30	1458

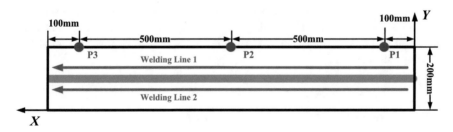

Fig. 5.3 Positions of measured points in welded T-joint

Table 5.3 Measured welding displacement of two typical fillet welded joints (unit: mm)

Welded joint	P1	P2	P3	P4	P5	P6
T200 fillet welded joint	2.0	2.1	2.3	0.2	0.0	0.2
T300 fillet welded joint	3.5	3.8	4.1	0.0	1.0	0.0

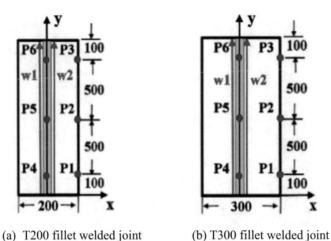

(a) T200 fillet welded joint   (b) T300 fillet welded joint

Fig. 5.4 Position of measured points in fillet welded joints (unit: mm)

Two fillet welded joints, both with 1200 mm in length and 9 mm in thickness but with different geometrical profiles, as shown in Fig. 5.4, were examined. One fillet welded joint had a web with 200 mm in height and a flange with 200 mm in width. This joint was identified as T200 fillet welded joint. The other joint was identified as T300 fillet welded joint owing to the 300 mm of web height and flange width. The welding conditions for these two typical welded joints are summarized in Table 5.2.

When the temperature of the welded joints completely cooled down to the ambient temperature, out-of-plane welding distortion was measured. Given that the examined welded specimen was highly symmetrical, Fig. 5.4 only shows the position of measured points over the right half part. The magnitudes of measured deflection are given in Table 5.3; note the increasing tendency of angular distortion along the welding line direction.

5.1.2 Thermal Elastic-Plastic FE Analysis

The wide application of computational approaches, in particular FE analysis, for solving engineering problems has paved the way for progressive acceptance and use of FE computation for welding simulations, such as transient temperature distribution, residuals stress, and welding distortion. Two fundamental approaches are typically used for welding distortion prediction [4]. One is thermal elastic-plastic FE analysis with solid-element model, in which welding is considered as a transient nonlinear problem. The other is elastic FE analysis with shell-element model, in which inherent strain/deformation is first evaluated and then applied to the welding line to represent mechanical behavior caused by arc heating. They both have advantages and limitations. The latter is superior in computing time but insufficient in terms of accuracy of the computed results because details of the welding condition may not be fully considered in some cases.

In this study, FE computation combined with the aforementioned analyses is proposed and employed. It consists of a transient thermal elastic-plastic FE analysis with iterative substructure method (ISM) to evaluate inherent deformation rapidly, an eigenvalue analysis for non-linear response when welding-induced buckling occurs, and elastic FE analysis applying inherent deformation as loading for welding distortion prediction. The procedure for this combined FE computation is as follows: inherent deformation is first previously evaluated through thermal elastic-plastic FE analysis of a typical welded joint with small size; then, this inherent deformation is employed for further investigation with eigenvalue analysis to calculate the critical load and possible buckling mode, and with elastic FE analysis to predict welding distortion in subsequent sections.

As discussed in Sect. 5.1.1, two typical fillet welded joints were made. According to the dimensions of the examined welded specimens, two FE models meshed with solid brick elements were created, as shown in Fig. 5.5. The total numbers of elements in T200 and T300 FE models were 30,100 and 28,700, respectively. The welding condition employed in the experiments was considered in the subsequent transient thermal elastic-plastic FE analysis. The arc heat efficiency was assumed to be 0.8 in the computation of the temperature field. For stress analysis, the rigid body motion was fixed to the boundary condition for obtaining the relative displacement.

An in-house FE code, which was implemented with the thermal elastic-plastic FE analysis mentioned above, was applied to solve temperature and mechanical problems during welding. Figure 5.6 individually shows the original and deformed shapes after welding of T200 and T300 fillet welded joints. Owing to the fixture of rigid body motion, evident angular distortion occurred in both T200 and T300 fillet welded joints, as shown in Fig. 5.6. When comparing the computed angular distortion with the corresponding experimental measurement, good agreement was

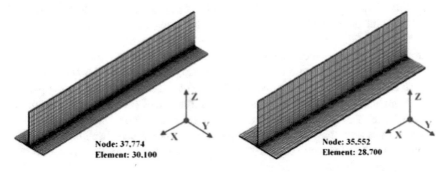

(a) FE model of T200 fillet welded joint   (b) FE model of T300 fillet welded joint

Fig. 5.5 Solid-element FE model of examined typical fillet welded joints

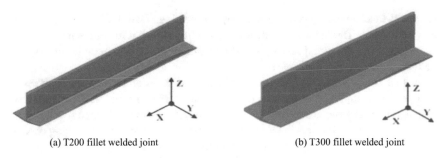

(a) T200 fillet welded joint   (b) T300 fillet welded joint

Fig. 5.6 Comparison of original and deformed shapes of fillet welded joints (scale: 5)

found, as shown in Fig. 5.7. Moreover, the changing tendency that angular distortion increased along the welding direction was observed from both experimental and computed results in the examined fillet welded joints.

5.1.3 FE Computation on Influence of Lateral Stiffener

To examine the influence of self-constraint on the prediction accuracy of out-of-plane welding distortion, a welded T-joint and an orthogonal stiffened welded structure as part of a typical ship panel were experimentally analyzed, including the geometrical profile, welding conditions, and welding distortion. Moreover, the influence of the lateral stiffener on FE computation was considered during the fabrication of a stiffened welded structure.

A solid-element model including the self-constraint supported by the lateral stiffener, as shown in Fig. 5.8, was also developed to re-evaluate the inherent deformation for actual fabrication. The welding conditions given in Table 5.4 were employed again for non-linear transient thermal and mechanical analyses. The boundary condition shown in Fig. 5.8 was still fixed to prevent rigid body motion. Figure 5.8 also shows the original and deformed shapes after welding with consideration of the self-constraint of the lateral stiffener. Figure 5.9 shows the computed contour of out-of-plane welding distortion. Note that non-uniform distribution of out-of-plane welding distortion appeared owing to the constraint of 2 lateral stiffeners.

The distributions of longitudinal and transverse plastic strains after cooling down are depicted in Fig. 5.10, in which the influence of 2 lateral stiffeners on the plastic strains can also be observed. To examine the influence of self-constraint supported by lateral stiffener on inherent deformation closely, comparisons among longitudinal inherent shrinkage, transverse displacement, and out-of-plane welding distortion along the welding line were conducted, as shown in Fig. 5.11 for the cases with and without lateral stiffeners.

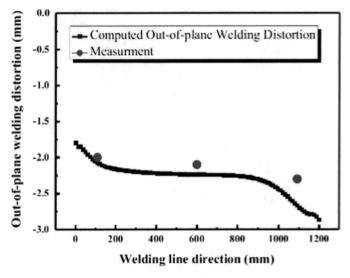

(a) T200 fillet welded joint

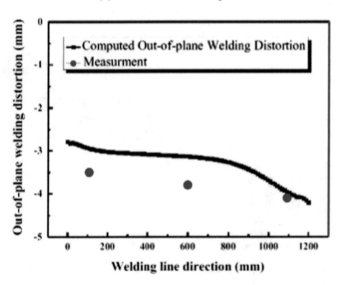

(b) T300 fillet welded joint

Fig. 5.7 Comparison of measured and computed angular distortions in fillet welded joints

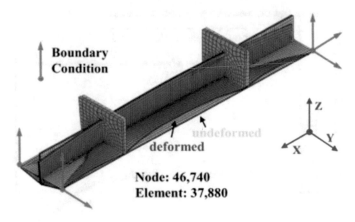

Fig. 5.8 Solid-element model of fillet welded joint with lateral stiffeners

Table 5.4 Welding conditions for fillet welding

Current (A)	Voltage (V)	Velocity (cm/min)
270	27	30

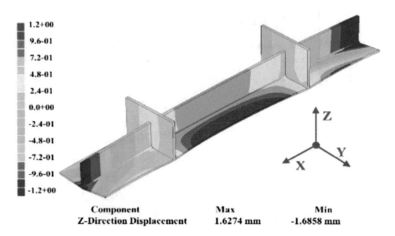

Component	Max	Min
Z-Direction Displacement	1.6274 mm	-1.6858 mm

Fig. 5.9 Contour of out-of-plane welding distortion with consideration of the influence of lateral stiffeners (deformed scale: 10)

Applying the same procedures for inherent deformation evaluation, the magnitudes of re-evaluated inherent deformation with consideration of the self-constraint of lateral stiffeners were obtained; they are summarized in Table 5.5. Note that a much less inherent bending was obtained with respect to the previously evaluated magnitude, as summarized in Table 5.6.

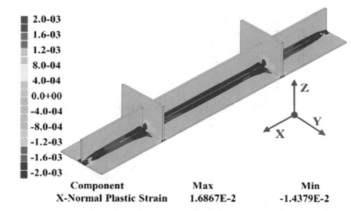

(a) Longitudinal plastic strains after cooling down

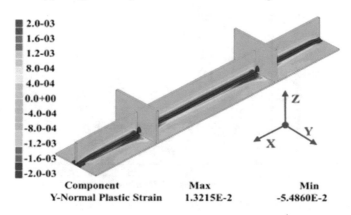

(b) Transverse plastic strains after cooling down

Fig. 5.10 Distributions of plastic strains with consideration of the influence of lateral stiffener

5.1.4 Inherent Deformation Evaluation

Inherent deformation has four basic components, namely the so-called longitudinal inherent shrinkage, transverse inherent shrinkage, longitudinal inherent bending, and transverse inherent bending. Usually, longitudinal inherent bending is neglected during elastic FE analysis owing to its much smaller magnitude and influence on welding distortion.

Concerning the other components, given that the longitudinal inherent shrinkage after welding has an extremely non-uniform distribution and concentrates close to the welding line owing to the strong self-constraint in the longitudinal direction, it is converted to tendon force. This force is applied to the welding line as a concentrated loading for representing the actual mechanical behavior in the longitudinal direction.

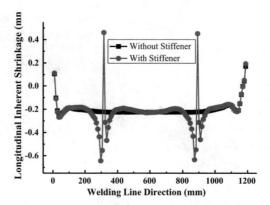

(a) Influence of lateral stiffener on longitudinal inherent shrinkage

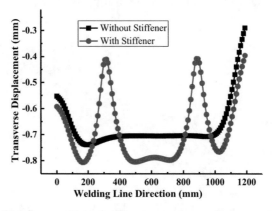

(b) Influence of lateral stiffener on transverse displacement

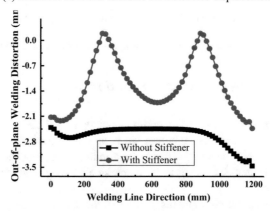

(c) Influence of lateral stiffener on out-of-plane welding distortion

Fig. 5.11 Comparison of mechanical response with consideration of the influence of lateral stiffeners

Table 5.5 Re-evaluated magnitudes of inherent deformation with consideration of the self-constraint of lateral stiffener

Longitudinal shrinkage force (kN)	Transverse inherent shrinkage (mm)	Plate bending (rad)
−301.787	0.692	0.00890

Table 5.6 Evaluated magnitudes of inherent deformation

Heat input (J/mm)	Longitudinal shrinkage force (kN)	Transverse inherent shrinkage (mm)	Plate bending (rad)	Stiffener bending (rad)
1495	−308.864	0.46675	0.04533	0.005768

By contrast, because of the weak self-constraint and its resulting mechanical behavior in the transverse direction, the transverse inherent shrinkage and inherent bending are transformed into displacement and bending radian, which can be evaluated directly from transverse displacement and angular distortion, respectively.

To evaluate the inherent deformation (tendon force) from plastic (inherent) strain based on its definition, longitudinal plastic strains after cooling down in T200 and T300 fillet welded joints were computed, as illustrated in Fig. 5.12 respectively. Neglecting the welding start and end, it is evident that a uniform distribution of longitudinal plastic strain near the welding line was obtained.

In general, inherent deformation has four components, as expressed in Eq. (2.8). They appear with different physical features owing to the different self-constraints supported by the surrounding base material [5]. Given that inherent deformation is the result of integrating inherent strain, and plastic strain is the dominant component of inherent strain after cooling down, computed plastic strains of examined welded T-joint in longitudinal and transverse directions presented clear differences for not only in terms of distribution but also in magnitude, as shown in Fig. 5.13. Their distributions on the middle cross-section perpendicular to the welding line are depicted in Fig. 5.14.

Longitudinal inherent shrinkage is converted to longitudinal shrinkage and tendon forces due to the fact that longitudinal inherent shrinkage after welding has an extremely non-uniform distribution and concentrates close to the welding line as a result of strong self-constraint in the longitudinal direction [2]. This shrinkage force was applied for the elastic FE analysis as an intrinsic loading to represent the actual mechanical response. To evaluate the longitudinal inherent shrinkage from plastic strain, computed longitudinal plastic strains after cooling down were obtained. They are shown in Fig. 5.13a. Neglecting the welding end effect, it is evident that a uniform distribution of longitudinal plastic strain close to the welding line occurred.

Because of the weak self-constraint in the transverse direction, the transverse inherent shrinkage and inherent bending are transformed into displacement and bending radian, which can be evaluated directly from transverse displacement and angular distortion, respectively. However, the longitudinal inherent bending is usually neglected during elastic FE analysis owing to its much smaller magnitude and

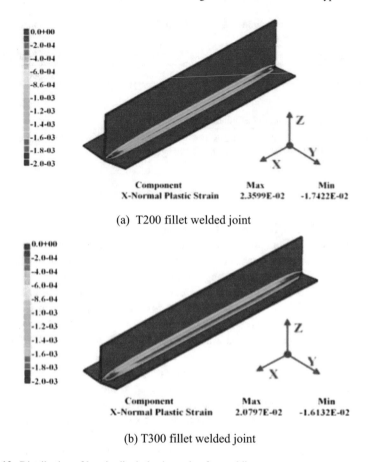

(a) T200 fillet welded joint

(b) T300 fillet welded joint

Fig. 5.12 Distribution of longitudinal plastic strain after welding

influence on welding distortion. The magnitudes for all the considered inherent deformation components are summarized in Table 5.7.

Therefore, the tendon force was evaluated through an integration approach by employing the computed longitudinal plastic strain, as shown in Fig. 5.12. This strain is always considered the dominant component of inherent strain. Transverse inherent shrinkage of plate and transverse inherent bending of plate and stiffener were evaluated with computed displacements, as shown in Fig. 5.14.

5.2 Welding Distortion of Stiffened Welded Structure

After the aforementioned investigation on fillet welded joints and the influence of lateral stiffeners, a stiffened welded structure of ship section was considered. Welding induced buckling was observed during experimental measurements . Then, an eigen-

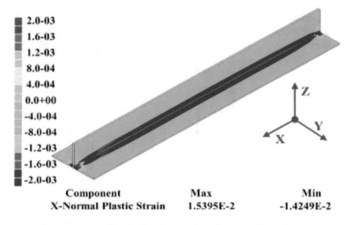

Component	Max	Min
X-Normal Plastic Strain	1.5395E-2	-1.4249E-2

(a) Longitudinal plastic strain after cooling down

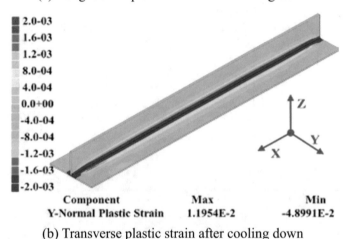

Component	Max	Min
Y-Normal Plastic Strain	1.1954E-2	-4.8991E-2

(b) Transverse plastic strain after cooling down

Fig. 5.13 Distributions of longitudinal and transverse plastic strains

value analysis was carried out to obtain the buckling mode and corresponding critical condition. Elastic FE analysis with welding inherent deformation was also carried out to predict out-of-plane welding distortion. Good agreement between computed results and corresponding measurements was obtained.

5.2.1 Fabrication of Orthogonal Stiffened Welded Structure

Figure 5.15 shows a mock-up of orthogonal stiffened welded structure as a part of a typical ship panel. This structure was made for experimentation. It was assembled by a square skin plate with 1200 mm in both length and width, two lower stiffeners

Fig. 5.14 Distributions of longitudinal and transverse plastic strains on middle cross-section

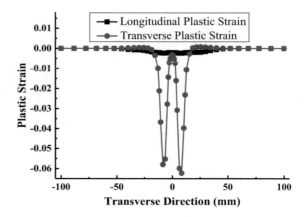

Table 5.7 Evaluated magnitudes of considered inherent deformation

Welded joint	Heat input (J/mm)	Tendon force (kN)	Transverse inherent shrinkage (rad)	Plate bending (rad)	Stiffener bending (mm)
T200 fillet welded joint	1495	308.864	0.46675	0.04533	0.005768
T300 fillet welded joint	1458	308.801	0.37176	0.04351	0.007884

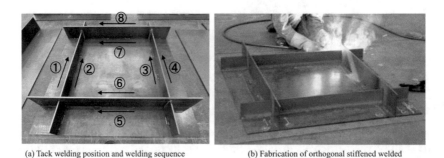

(a) Tack welding position and welding sequence   (b) Fabrication of orthogonal stiffened welded structure with welding

Fig. 5.15 Overview of orthogonal stiffened welded structure

with 100 mm in height, and two higher stiffeners with 150 mm in height. In addition, all stiffeners had 1200 mm in length and 9 mm in thickness. The welding conditions listed in Table 5.4 were employed.

Positions of the measured points on the skin plate of the examined orthogonal stiffened welded structure are shown in Fig. 5.16. These points were considered to

Fig. 5.16 Measured points in orthogonal stiffened welded structure

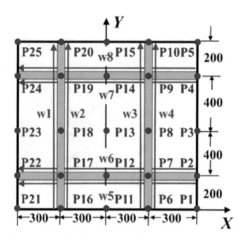

measure the out-of-plane welding distortion. The measured magnitudes of out-of-plane welding distortion are summarized in Table 5.8. Note from these results that the center point has an upward displacement with slightly larger magnitude, and edge parts bended downward to the opposite direction with respect to the center region deflection. This type of deformed shape had to be caused by the buckling behavior. Moreover, welding induced buckling was generated during the fabrication of the orthogonal stiffened welded structure.

Table 5.8 Measured welding displacement of cross stiffened welded structure (unit: mm)

Measurement 1 in parallel stiffened welded structure				
$Z(P25) = -1.3$	$Z(P20) = -0.3$	$Z(P15) = 0.5$	$Z(P10) = -0.5$	$Z(P5) = -0.1$
$Z(P24) = -1.1$	$Z(P19) = 0.7$	$Z(P14) = 1.8$	$Z(P9) = 0.3$	$Z(P4) = -1.7$
$Z(P23) = -3.4$	$Z(P18) = 0.5$	$Z(P13) = 9.7$	$Z(P8) = 1.0$	$Z(P3) = -1.9$
$Z(P22) = -1.7$	$Z(P17) = 0$	$Z(P12) = 1.5$	$Z(P7) = 0.7$	$Z(P2) = -2.4$
$Z(P21) = -1.7$	$Z(P16) = -1.4$	$Z(P11) = -0.4$	$Z(P6) = -1.0$	$Z(P1) = -0.7$
Measurement 2 in parallel stiffened welded structure				
$Z(P25) = 0.5$	$Z(P20) = -1.0$	$Z(P15) = -2.0$	$Z(P10) = 0.5$	$Z(P5) = 0.5$
$Z(P24) = 0$	$Z(P19) = 0$	$Z(P14) = -0.5$	$Z(P9) = 0$	$Z(P4) = -0.5$
$Z(P23) = 0$	$Z(P18) = 0.5$	$Z(P13) = 7.0$	$Z(P8) = -0.5$	$Z(P3) = -1.0$
$Z(P22) = 0$	$Z(P17) = 0$	$Z(P12) = 0$	$Z(P7) = -0.5$	$Z(P2) = -1.5$
$Z(P21) = -1.0$	$Z(P16) = -1.0$	$Z(P11) = -3.0$	$Z(P6) = -1.5$	$Z(P1) = -1.5$

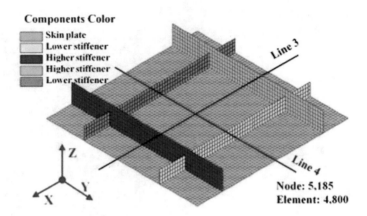

Fig. 5.17 Shell-element models of examined cross stiffened welded structure

5.2.2 Elastic FE Analysis with Shell-Element Model

For the examined parallel and cross stiffened welded structures, not only the general welding distortion but also buckling deformation may be produced by the applied welding procedure after cooling down in case of usage of relatively thin plates.

Given that welding induced buckling is a stable problem of non-linear response, it is difficult to predict the magnitude and deformed shape of buckling distortion. Thus, the large deformation theory expressed in Eq. (2.9) should be considered. Eigenvalue analysis was first carried out for investigating whether buckling behavior occurred in parallel and cross stiffened welded structures. The evaluated in-plane inherent shrinkages reported in the previous section, which are considered the dominant reason for welding induced buckling, were employed as loading in such eigenvalue analysis [3]. Subsequently, all the components of inherent deformations are employed in an elastic FE analysis to predict out-of-plane welding distortion including bending and welding buckling.

FE meshes with shell elements for the examined parallel and cross stiffened welded structures are shown in Fig. 5.17. They were employed for the eigenvalue and elastic FE analyses jut mentioned above. Interface elements were automatically created between adjacent components with different colors for representing the welding joint, and inherent deformations caused by the corresponding welding were applied on the interface elements to model the mechanical behavior, e.g., welding shrinkage and out-of-plane welding distortion.

5.2.2.1 Eigenvalue Analysis for Buckling Behavior

Given that welding induced buckling is a non-linear response during fabrication, eigenvalue analysis with the FE model shown in Fig. 5.17 was conducted to examine whether buckling behavior occurred. The computed results of eigenvalue analysis

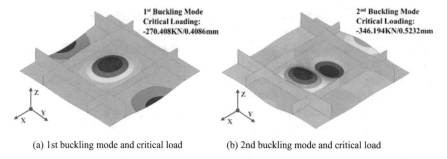

(a) 1st buckling mode and critical load   (b) 2nd buckling mode and critical load

Fig. 5.18 Two possible buckling modes of orthogonal stiffened welded structure

included the two lowest possible buckling modes of orthogonal stiffened welded structure under current welding condition, as shown in Fig. 5.18.

To consider buckling behavior, in particular welding induced buckling, eigenvalue analysis is a preferred computational approach. Given welding specifications (in-plane inherent shrinkages) and examined welded structure, eigenvalue analysis can produce the magnitude of critical condition and corresponding buckling mode.

Applying the magnitudes of in-plane inherent shrinkages summarized in Table 5.7 gave rise to computed results that included the two lowest possible buckling modes of examined different stiffened welded structures under current welding condition, as shown in Fig. 5.18.

From the calculated eigenvalue parameter of each buckling mode, the lowest buckling mode shown in Fig. 5.18 occurred in the examined cross stiffened welded structure under the considering welding procedure, in which the calculated eigenvalue parameter was below 1.0. This means that the applied in-plane inherent shrinkages were greater than the required loading of the lowest buckling mode. Specifically, 87.56% of the applied in-plane inherent shrinkages generated buckling deformation.

$$eigenvalue \ parameter = \frac{n\text{th } critical \ loading}{applied \ loading} \tag{5.1}$$

From the calculated eigenvalue parameter defined by Eq. (5.1) for each buckling mode, the lowest buckling mode shown in Fig. 5.18a appeared during the fabrication of the examined orthogonal stiffened welded structure with respect to thermal loading and geometrical configuration. The calculated eigenvalue parameter of the 1st buckling mode was 0.8756. The buckling critical value was below 1.0 as well. It means that the applied welding thermal loading (in-plane component of welding inherent deformation) was greater than the lowest required loading of buckling occurrence. For the 2nd buckling mode, it required much greater magnitudes of longitudinal shrinkage force and transverse shrinkage. This type of buckling mode did not occur in the conducted welding experiments.

5.2.2.2   No Consideration of the Influence of Lateral Stiffeners

The shell-element models shown in Fig. 5.17 was also used for elastic FE computation. The evaluated inherent deformation is summarized in Table 5.7. It was applied to the welding line for welding mechanical analysis. The boundary condition was still fixed to preclude rigid body motion.

Owing to the different out-of-plane welding distortion patterns, parallel and cross stiffened welded structures were respectively investigated with the large deformation theory. The results are presented next.

Concerning eigenvalue analysis, buckling behavior occurred in the production of the examined cross stiffened welded structure. It was necessary to consider the large deformation theory in FE computation.

All the inherent deformations evaluated from analyses of two typical fillet welded joints were employed. Figure 5.19 shows the overview of out-of-plane welding distortion distribution. Note that all the edges and central region bended upward with respect to the longitudinal and transverse stiffeners. There was no supporting evidence of buckling behavior from the computed results. Thus, it is difficult to conclude whether this examined cross stiffened welded structure buckled or not. A possible reason for this is the inherent bending, which was evaluated with a relatively large magnitude from fillet welding. However, the previous considered fillet welding under rigid body motion could freely bend without other constraints supported by stiffeners. Actually, the skin plate in the examined cross stiffened welded structure was strongly fixed to prevent bending by longitudinal and transverse stiffeners. Therefore, less magnitude of inherent bending should be applied for more rational consideration according to the research by Ma et al. [6].

According to the above eigenvalue analysis, welding induced buckling with the lowest buckling mode was generated during the fabrication of the orthogonal stiffened

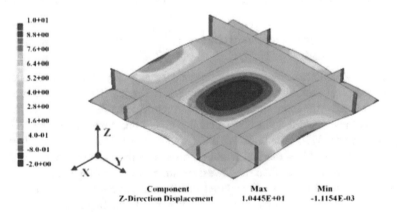

Fig. 5.19 Computed out-of-plane welding distortion of examined cross stiffened welded structure with all inherent deformation components (scale: 5)

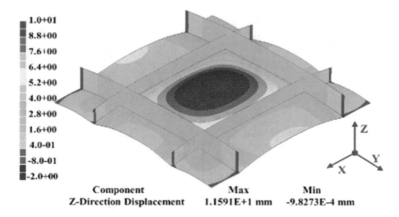

Component	Max	Min
Z-Direction Displacement	1.1591E+1 mm	-9.8273E-4 mm

Fig. 5.20 Computed out-of-plane welding distortion without considering the influence of lateral stiffeners (deformed scale: 10)

welded structure. It was necessary to consider the finite strain theory for the later elastic FE computation.

Applying the magnitudes of inherent deformation listed in Table 5.6 without considering the influence of self-constraint supported by lateral stiffener, elastic FE analysis was carried out with the shell-element model shown in Fig. 5.17 and finite strain theory for out-of-plane welding distortion prediction. Figure 5.20 shows the overview of out-of-plane welding distortion distribution. Note that all the edges and center region bended upward with respect to the original shape. There was no sufficient evidence showing the occurrence of buckling from this computed result, so it was difficult to infer whether this orthogonal stiffened welded structure buckled or not.

A comparison of computed out-of-plane welding distortion and corresponding measurement is shown in Fig. 5.21. A much larger difference of out-of-plane welding distortion can be observed. The potential explanation for this is inherent bending, which was evaluated directly from TEP FE computation without considering the self-constraint supported by lateral stiffeners.

The generation mechanism of welding induced buckling can be explained as follows. In-plane inherent shrinkages are the dominant cause that determine the occurrence of buckling behavior. Inherent bending with other geometrical imperfections influences the magnitude of out-of-plane welding distortion. This may sometimes obscure the buckling behavior owing to its large magnitudes [7]. Thus, self-constraint supported by not only the surrounding base material but also lateral stiffeners should be considered to evaluate inherent deformation accurately for out-of-plane welding distortion prediction. In particular, self-constraint supported by the surrounding base material sets inherent deformation as individual physical representation, whereas self-constraint supported by lateral stiffeners significantly influences the magnitude of bending component and the final deformed shape.

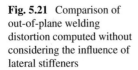

Fig. 5.21 Comparison of out-of-plane welding distortion computed without considering the influence of lateral stiffeners

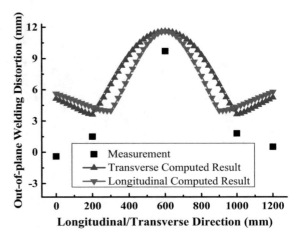

5.2.2.3   Consideration of the Influence of Lateral Stiffener

As discussed in the previous section, self-constraint of lateral stiffeners significantly influences the magnitude of inherent bending, which in turn eventually influences the final deformed shape, the magnitude of out-of-plane welding distortion, and the computational prediction of buckling behavior.

Considering the self-constraint supported by lateral stiffeners, the magnitude of inherent deformation in Table 5.5 was employed as welding loading for elastic FE analysis. Figure 5.22 depicts an identical deformed shape as the one from eigenvalue analysis shown in Fig. 5.18a. Figure 5.22 also shows the contour of computed out-of-plane welding distortion, in which the center region deforms upward and all the edges

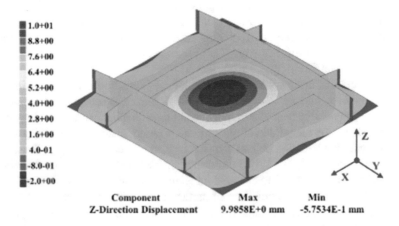

Fig. 5.22 Computed out-of-plane welding distortion with consideration of the influence of lateral stiffener (deformed scale: 10)

Fig. 5.23 Comparison of welding distortion computed with consideration of the influence of lateral stiffeners

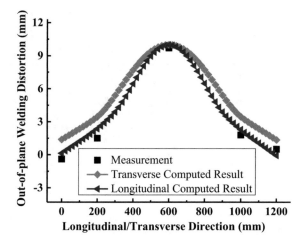

deform downward with respect to the original shape. From this type of out-of-plane welding distortion distribution, we can confirm with certainty that this orthogonal stiffened welded structure buckled during welding fabrication. The downward out-of-plane welding distortion at edges had to be caused by welding buckling.

To confirm the accuracy of the predicted out-of-plane welding distortion, the out-of-plane welding distortions between elastic FE computation and corresponding measurements were compared. Figure 5.23 shows that the computed out-of-plane welding distortion presented much better agreement with the corresponding measurement. Clearly, both sides of the stiffeners underwent different deformed directions due to welding induced buckling.

5.2.2.4   Consideration of Only In-Plane Inherent Shrinkage

We can conclude from the above discussion that, owing to the strong fixture of all stiffeners on skin plate bending during welding, in-plane inherent shrinkages and a less inherent bending must be applied for elastic FE analysis. A similar deformed shape to that shown in Fig. 5.18a from eigenvalue analysis was obtained, as shown in Fig. 5.24. This figure also shows the computed results, where the central region deformed upward and all the edges deformed downward with respect to the original shape. From this type of out-of-plane welding distortion distribution, we can confirm with certainty that this cross stiffened welded structure buckled under welding. The downward out-of-plane welding distortion at edges had to be caused by welding induced buckling. A comparison of deflections between computed and measured results was also performed. Figure 5.25 shows the computed distribution of deflections after cooling down points on lines 3 and 4, as shown in Fig. 5.17. They were in good agreement with those at the measured points. Clearly, both sides of the stiffeners underwent different deformed directions due to welding buckling.

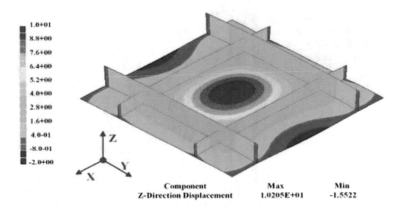

Fig. 5.24 Computed out-of-plane welding distortion of examined cross stiffened welded structure with appropriate inherent bending (scale: 5)

5.3 Conclusions

As a summary, out-of-plane welding distortions of a welded T-joint and an orthogonal stiffened welded structure as a part of a typical ship panel were examined through experiments and FE computations. It was necessary to consider the self-constraints supported by the surrounding base material and lateral stiffeners for out-of-plane welding distortion prediction. Welding induced buckling was predicted with good agreement with the corresponding measurements. From the computational procedure and computed results, it was confirmed that elastic FE computation is an ideal and practical computational approach for welding distortion prediction in large and complex welded structures. The following conclusions were also drawn:

(1) Thermal elastic-plastic FE computation with in-house code was carried out to predict angular distortion of fillet welded joint, with good agreement with the corresponding measurements.

(2) A combined computational approach is proposed, in which thermal elastic-plastic FE analysis is carried out for inherent deformation evaluation, eigenvalue analysis is employed for calculation of the critical buckling condition under welding, and elastic FE analysis is applied for welding distortion prediction.

(3) Eigenvalue analysis with shell-element model and inherent deformation is efficient for investigation on welding induced buckling. It can provide each critical buckling condition and corresponding buckling mode.

(4) Considering the self-constraint supported by the surrounding base material and lateral stiffeners, both the deformed shape and magnitude of out-of-plane welding distortion in the computed results with elastic FE analysis presented consistent features with measurements.

(5) Concerning the mechanism, in-plane inherent shrinkages are the dominant cause of welding induced buckling. Inherent bending triggers the occurrence

Fig. 5.25 Comparison of computed and measured results for out-of-plane welding distortion of examined cross-stiffened welded structure

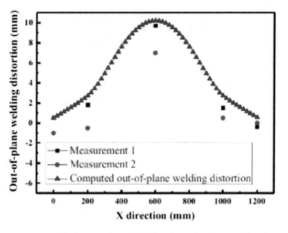

(a) Comparison of deflection of points on line 3

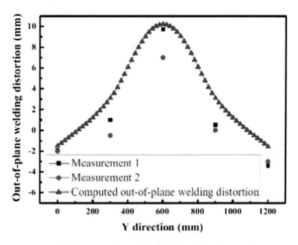

(b) Comparison of deflection of points on line 4

of welding buckling when the critical buckling condition is reached. However, buckling behavior may occur due to inherent bending with large enough magnitude.

References

1. Masubuchi K (1980) Analysis of welded structures: residual stresses, distortion and their consequences. Pergamon Press, Oxford
2. Wang JC, Ma N, Murakawa H (2015) An efficient FE computation for predicting welding induced buckling in production of ship panel structure. Mar Struct 41:20–52

3. Wang JC, Rashed S, Murakawa H (2014) Mechanism investigation of welding induced buckling using inherent deformation method. Thin-Walled Struct 80:103–119
4. Murakawa H, Deng D, Ma N (2010) Concept of inherent strain, inherent stress, inherent deformation and inherent force for prediction of welding distortion and residual stress. In: Proceedings of the international symposium on visualization in joining and welding science, Nov 11–12, Osaka, Japan, pp 115–116
5. Moshaiov A, Song H (1990) Near- and far-field approximation for analyzing flame heating and welding. J Therm Stresses 13(1):1–19
6. Ma N, Huang H, Murakawa H (2015) Effect of jig constraint posi- tion and pitch on welding deformation. J Mater Process Technol 221:154–162
7. Wang JC, Rashed S, Murakawa H, Luo Y (2013) Numerical prediction and mitigation of out-of-plane welding distortion in ship panel structure by elastic FE analysis. Mar Struct 34:135–155

Chapter 6
Application of Computational Welding Mechanics for Accurate Fabrication of Ship Structure

The investigation on fillet welded joints and stiffened welded structures above described examined out-of-plane welding distortion behaviors through experiments and advanced FE computations to control fabrication accuracy and clarify the generation mechanism. Moreover, typical ship structures including hatch coaming structures of bulk-cargo ship, ship panels, watertight transverse bulkhead structures, and torsional box structures of container ships were sequentially examined by means of elastic FE analysis with welding inherent deformation. In particular, when thin plate sections are designed and fabricated, not only the conventional welding distortion but also welding induced buckling is generated. Welding induced buckling results in loss of dimensional control and structural integrity, and may delay the fabrication schedule and increase the fabrication cost when mitigation is carried out [1]. During the correction, it is difficult to reduce the welding induced buckling completely owing to its features of diversity and instability. Therefore, it is better to avoid the generation of welding induced buckling whenever possible.

6.1 Welding Distortion Reduction for Hatch Coaming Production

Hatch coaming is the vertical wall surrounding an opening in a ship deck. It provides a frame on top of which the hatch cover is fitted. In conjunction with the hatch cover, it prevents entry of water into ship holds [2]. Figure 6.1 shows a hatch coaming and a folding hatch cover. Hatch coaming structures have been widely used on ships for a long time. It is necessary to avoid misalignment between the hatch coaming top surface and the hatch cover to maintain water tightness and to allow smooth mechanical operation of the hatch cover. Actually, this underlying problem of misalignment between the hatch coaming top surface and hatch cover is the result of the welding process during hatch coaming production. Welding is commonly used because of its

H. ZHOU and J. WANG, *FE Computation on Accuracy Fabrication of Ship and Offshore Structure Based on Processing Mechanics*, https://doi.org/10.1007/978-981-16-4087-2_6

Fig. 6.1 Hatch coaming and
folding hatch cover

high productivity and flexible practice. However, welding distortion always occurs
due to the non-uniform expansion and contraction of the weld and surrounding base
material [3]. Welding distortion influences the assembly accuracy of welded structures and, when mitigated, it delays the production schedule and increases the fabrication cost [4]. Lee et al. [5] studied the control method of global bending distortion
resulting from the fabrication process of the hatch cover in a container ship. To this
end, a three-dimensional measurement instrument was used to measure the transi-
tional behavior of global bending distortion. The conclusion was that the cause of this
distortion is the bending moment associated with the longitudinal shrinkage force
and the transverse shrinkage resulting from welding and flame heating.

In a conventional assembly process, the hatch coaming is divided into several
parts that are sequentially assembled to the deck and welded with adjacent parts. This
process is time-consuming because of multiple lifting and fitting of hatch coaming
parts. Figure 6.2 shows a hatch coaming consisting of 16 parts depicted with different
colors assembled to the ship deck (1: ship deck; 2–17: parts of the hatch coaming). To
improve the production efficiency, an enhanced assembly process for hatch coaming
production is proposed. The complete hatch coaming is assembled on the shop floor
and then lifted as one part, fitted, and welded to the deck. This process is illustrated
in a simple way in Fig. 6.3 (1: ship deck; 2: complete hatch coaming).

Using the improved assembly process for hatch coaming production, the time
consumption can be significantly reduced according to the measured database of
a shipyard (Nankai Shipyard, Japan), as shown in Table 6.1. However, when the
already assembled complete hatch coaming is fitted to the deck, a varying gap is
usually found between the lower side of the hatch coaming and the deck because
of the welding distortion produced during assembling the hatch coaming on the
shop floor. Longitudinal and transverse shrinkage deformations induced by welding
between the hatch coaming and ship deck are influenced by the width of this gap.

These shrinkage deformations influence the dimensional accuracy and alignment
between the hatch coaming and hatch cover. The target of our research was the
reduction of welding distortion of hatch coaming top surface during the improved

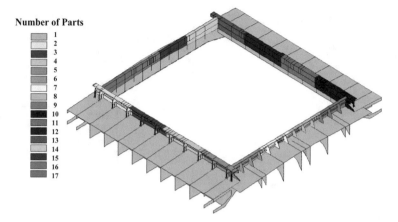

Fig. 6.2 Conventional assembly process for hatch coaming production

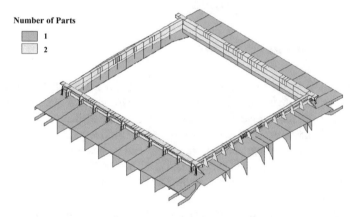

Fig. 6.3 Improved assembly process for hatch coaming production

assembly process for hatch coaming production. In this context, improvement of transverse shrinkage of multi-pass fillet welded joints with different gaps and welding pass sequences was first addressed through thermal elastic-plastic finite element analysis. Inherent longitudinal and transverse shrinkage deformations were evaluated for each gap and each welding pass sequence. Applying these inherent deformations, elastic analysis was used to predict the welding distortion of the hatch coaming, which is influenced by the gap and welding pass sequence, and assess the alignment between the hatch coaming top surface and hatch cover.

Table 6.1 Comparison of time consumption between assemble processes

Conventional assembly process	Time consumption (h)	Improved assembly process	Time consumption (h)
Lifting and fitting process	96	Parts of hatch coaming	31
		Hatch coaming and deck	87
Welding process	278	Parts of hatch coaming	34
		Hatch coaming and deck	138
Total consumed time	374		290

6.1.1   Experimental Procedure

First, experiments with three specimens that modeled the welded joint between the hatch coaming and deck with different gaps were carried out. Figure 6.4 shows the dimensions of the studied welded joint; clamping is also displayed. The length and width of the flange were 1000 mm and 300 mm, respectively; the height of the web was 300 mm; and the thicknesses of the flange and web were 16 mm and 12 mm, respectively. The arrangements and sequence of the welding pass are illustrated in Fig. 6.5 for specimens with different gaps. The heat inputs of each welding pass are listed in Table 6.2.

The measurement equipment for welding distortion was fixed on the web, as shown in Fig. 6.6. As a result of the changing angle between the web and flange, the measurement equipment rotated together with the web. The changed distance

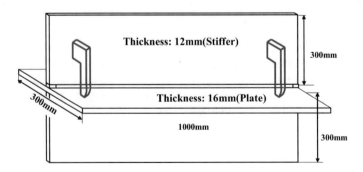

Fig. 6.4 Dimensions of specimen for experimentation

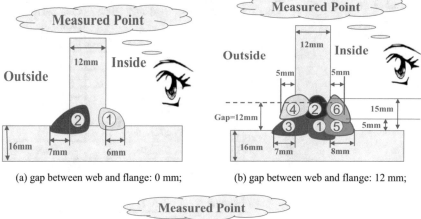

(a) gap between web and flange: 0 mm;   (b) gap between web and flange: 12 mm;

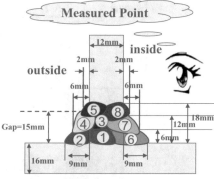

(c) gap between web and flange: 15 mm

Fig. 6.5 Arrangement of welding pass for specimens with different gaps

between the measurement equipment and the flange after each welding pass for different specimens is also given in Table 6.2.

6.1.2 Evaluation of Inherent Deformation of Fillet Welded Joints

To predict welding distortion by elastic FE analysis, it was necessary to evaluate the inherent deformation previously. First, thermal elastic-plastic (TEP) FE analysis was used to predict the welding distortion of the three fillet welded specimens with different gaps. Computational results were compared with the corresponding experimental measurements to verify the analytical approach. Then, the influence of welding pass sequence on web transverse shrinkage of two specimens with different gaps was investigated through the computational approach. Improved welding pass sequences are proposed to reduce web transverse shrinkage. Inherent deformation

Table 6.2 Heat input and changed distance for each welding pass of the three specimens

Gap	Welding pass	Heat input (J/mm)	Changed distance (mm)
0 mm	First	1404	−1.037559
	Second	2685	−0.637011
12 mm	First	1030.63	−0.525002
	Second	1167.34	−0.93255
	Third	2410.12	−0.52264
	Fourth	3021.13	−0.57105
	Fifth	3060.08	−1.65963
	Sixth	3431.47	−2.90881
15 mm	First	900.13	−0.11432
	Second	926.2	−0.37234
	Third	2359.06	−1.21282
	Fourth	2522.43	−0.87721
	Fifth	2630.69	−0.52251
	Sixth	2650.98	−1.73698
	Seventh	2653.94	−2.45843
	Eighth	2484.32	−3.46163

Fig. 6.6 Specimen and measurement equipment

was evaluated from the computed results for the original and improved welding pass sequences and can be used in subsequent elastic FE analysis.

6.1.2.1 Verification of Thermal Elastic-Plastic FE Analysis

Considering the same welding conditions of the corresponding experimental specimens, three-dimensional finite element analyses were carried out using the TEP FE approach outlined previously. Figure 6.7 shows the solid-element model used in our analysis. Boundary conditions that restrain only the rigid body motion of the specimen were used, as shown in Fig. 6.7.

Because of the location of the measurement equipment and welding pass sequence, the measured results comprised two parts: transverse shrinkage of the web and a shortening caused by angular distortion (rotation). For this complex and combined welding distortion in the fillet welded joint, the distance between the measurement equipment and the flange of the fillet welded joint was measured to investigate this special combined welding distortion (transverse shrinkage and angular distortion). Note that during the welding process, the web transverse shrinkage always increased with each welding pass. Comparisons between computed and measured changed distances between the measurement equipment and the flange of the three specimens are shown in Fig. 6.8. Note that the computed changed distances are in good agreement with the corresponding measurements. As a result of the experiment, the sign of changed distance is deformed tendency such as minus means that the measuring equipment is close to the flange of fillet welded joint and absolute value of changed distance is deformed magnitude. Because of the angular distortion caused by the opposite welding line, the web of the fillet welded joint with fixed measurement equipment rotated toward the opposite side of the fillet welded joint. This made the magnitude of changed deformation became smaller than that after the previous welding line, regardless of the transverse shrinkage of the web.

Taking the specimen with a gap of 12 mm as an example, the magnitude of measured changed distance increased after passes 1 and 2. When the welding passes 3 and 4 were applied, the web together with the measurement equipment rotated to the outside, which is the opposite side of the measurement equipment. The angular distortion reduced the magnitude of changed distance, but it was not large enough to change the deformed tendency (also minus). The web transverse shrinkage was also

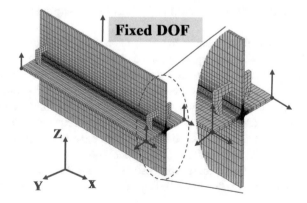

Fig. 6.7 Solid-element model for thermal elastic-plastic FE analysis

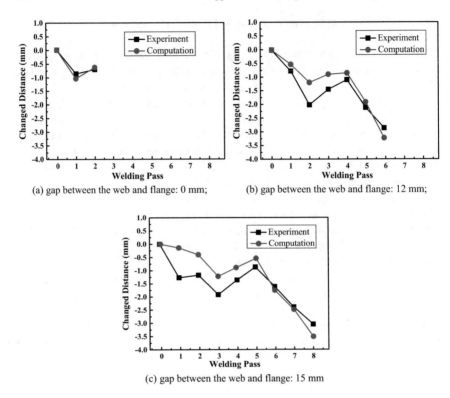

(a) gap between the web and flange: 0 mm;   (b) gap between the web and flange: 12 mm;

(c) gap between the web and flange: 15 mm

Fig. 6.8 Comparison between experimental and computational results

increased with these two passes. Finally, welding passes 5 and 6 located inside, where the measurement equipment was fixed, were applied. The magnitude of changed distance increased very quickly because of the combined contribution of transverse shrinkage and angular distortion.

6.1.2.2 Improvement of Welding Pass Sequences

In the particular hatch coaming under study, longitudinal shrinkage may not cause an appreciable distortion of the hatch coaming top surface. This is because of the high stiffness of the structure and because welding was carried out close to the neutral axis of the combined hatch coaming and deck structure. Therefore, in this study, attention was focused on web transverse shrinkage and how to reduce it during the assembly process. The welding pass sequence was considered to achieve this reduction. Computational approaches were used to predict web transverse shrinkage when the improved pass sequences were adopted.

The welded joints with gaps of 12 and 15 mm were investigated. The web and flange had to be connected as late as possible because the web transverse shrinkage

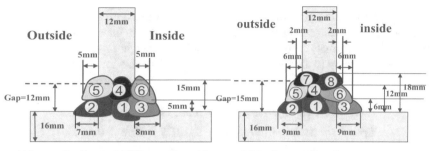

(a) Improved sequence of welding pass for multi-pass   (b) Gap between the web and flange: 15 mm

Fig. 6.9 Image of welding bead arrangement for fillet welded joints

was produced only during and after the web and flange were first connected by welding. The improved welding pass sequences reflecting this logic are shown in Fig. 6.9. Welding passes were arranged on the flange in such a way that the gap between the web and the flange was first decreased and then the web and flange became connected together.

When the welding pass sequences were exclusively considered, the same heat input for each welding pass was used. The web transverse shrinkage for different gaps on fillet welded joints was predicted by using TEP FE analysis using the improved welding pass sequences. Then, the computed results were used to evaluate the inherent deformation for elastic analysis.

6.1.2.3 Evaluation of Inherent Deformation

Computational results obtained by TEP FE analysis were used to evaluate the inherent deformation. According to the definition of inherent deformation, inherent longitudinal and transverse shrinkage deformations of specimens with different gaps and different welding pass sequences were obtained, as reported in Table 6.3. Compared with the current welding pass sequences, the improved welding pass sequences were found to reduce the inherent longitudinal and transverse shrinkage deformations significantly. Inherent bending deformation had only local effects and could be ignored during the elastic FE analysis to predict welding distortion of the hatch coaming.

Table 6.3 Inherent shrinkage deformations of different gaps considering the welding pass sequence

Case (gap) (mm)	Current welding sequence		Improved welding sequence	
	Longitudinal shrinkage	Transverse shrinkage	Longitudinal shrinkage	Transverse shrinkage
0	0.303	0.42857	0.303	0.42857
12	0.327	0.92421	0.322	0.43007
15	0.337	1.02687	0.404	0.7193

Using the inherent deformations in Table 6.3 and based on the assumption that such deformations have linear relationships with gap width, linear functions relating the inherent deformation to gap width for different welding pass sequences were evaluated according to the following expression, which makes use of the linear least squares method:

$$Long : shrinakage(current) = 0.00219 \times Gap + 0.30262$$
$$Tran : shrinakage(current) = 0.04029 \times Gap + 0.43059$$
$$Long : shrinakage(improved) = 0.00526 \times Gap + 0.29564$$
$$Trans : shrinakage(improved) = 0.01388 \times Gap + 0.40105 \qquad (6.1)$$

6.1.3 Prediction of Welding Distortion of Hatch Coaming Using Elastic FE Analysis

When the inherent deformations of the hatch coaming-to-deck fillet welded joint were predicted for different gaps, elastic FE analysis was used to investigate welding distortion of the hatch coaming by applying the inherent deformations to the weld lines. Figure 6.10a shows the shell-element model of the hatch coaming and deck structure in which the welding line was divided into 16 parts to consider the varying gap between the hatch coaming and the deck at different locations. The gaps at different parts are also shown in Fig. 6.10a. The boundary condition shown in Fig. 6.10b is a reasonable approximation of the restraint provided by the surrounding deck structure and was used in the elastic FE analysis. To examine the misalignment between the hatch coaming and hatch cover closely, as well as the influence by welding distortion of hatch coaming, lines 1 and 2 on the top surface of the hatch coaming, shown in

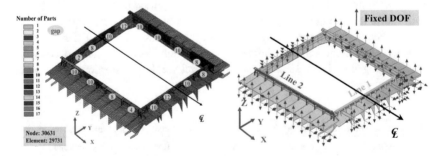

(a) Shell-element model of hatch coaming and deck structure; (b) Boundary condition for elastic FE analysis

Fig. 6.10 Shell-element model and boundary condition

Fig. 6.10b, were selected to compare the vertical displacement (Z direction) using different welding pass sequences.

Inherent deformations were applied to the weld lines between the hatch coaming and deck structure. Figure 6.11 shows the vertical displacement of hatch coaming considering the different welding pass sequences. An evident difference of vertical displacement of hatch coaming was observed using different welding pass sequences.

Figure 6.12 shows a comparison of vertical displacement of hatch coaming top surface on lines 1 and 2 shown in Fig. 6.10 for different welding pass sequences. Vertical displacement of hatch coaming top surface was significantly reduced with respect to the current welding pass sequence when the improved welding pass sequence was used.

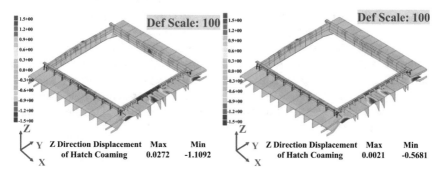

(a) Current welding multi-pass sequence   (b) Improved welding multi-pass sequence

Fig. 6.11 Vertical displacement of hatch coaming predicted by elastic FE analysis

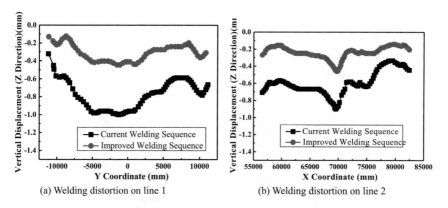

(a) Welding distortion on line 1   (b) Welding distortion on line 2

Fig. 6.12 Comparison of welding distortion between different welding pass sequences

6.2 Investigation on Welding Induced Buckling for Ship Panel Fabrication

Elastic FE analysis with inherent deformation method is an ideal and advanced computational approach for predicting welding induced buckling in which the large deformation theory is employed to consider geometrical non-linear response. In this study, the implementation of an elastic FE analysis to predict welding distortion is comprehensively introduced, and its application on welding induced buckling of ship panel structures is demonstrated. First, a fillet welded joint that exists in the considered ship panel was performed with sequential welding, and welding angular distortion was measured. A TEP FE analysis of this examined fillet welded joint was carried out. The computed results were validated through experiments by comparing the welding angular distortion. Then, the inherent deformation produced by sequential welding applied to the assembly fillet welded joint was evaluated. Simultaneous welding of the fillet welded joint was numerically examined and its inherent deformation was also evaluated. To predict the welding induced buckling of an actual ship panel, a part of ship panel was selected and examined by elastic FE analysis. The results were compared with the classical theoretical solution. Evident welding bucking in the ship panel structure was observed in the computed results considering the large deformation theory.

6.2.1 Experimental Procedures and Measurement

A typical fillet welded joint, shown in Fig. 6.13, was first selected as a test model in this study. This fillet welded joint was assembled by two components: a flange with dimensions 300 mm × 300 mm × 6 mm and a web with dimensions 300 mm × 100 mm × 6 mm. In this experiment, the CO_2 arc welding process was carried out and two welding passes were sequentially performed. The welding conditions are shown in Table 6.4. After welding, the angular distortion of one edge was measured.

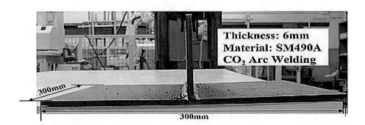

Fig. 6.13 Front view of the experimental fillet welded joint

Table 6.4 Welding conditions of fillet welded joint	Current (A)	Voltage (V)	Velocity (mm/s)	Efficiency
	170–190	20–30	4.0–5.0	0.7–0.8

6.2.2 Thermal Elastic-Plastic FE Computation of Fillet Welding

Inherent deformation is the basis of elastic FE analysis. Its magnitude has a large influence on the deformed shape of the welded structure and the final predictive accuracy of welding distortion. According to previous discussions, the inherent deformations defined in Eq. (2.8) were determined by the inherent strain caused by heating and cooling processes during welding, which mostly depend on the welding heat input, material properties, and thickness of the welded joint when the welded structure is large enough. Furthermore, for a long welding line, constant magnitudes of inherent deformations can be evaluated through the observation of uniform distributions in the middle region of the welding line while ignoring the end effect.

Generally speaking, there are two procedures to evaluate the magnitudes of inherent deformations. The first one is inverse analysis based on welding distortion experimentally measured or computed by FE analysis. In this case, inherent deformations are inversely calculated using displacements of some critical points of the welded joint [6–8]. The second procedure consists of integrating the inherent strains (mostly plastic strain) according to the definition of inherent deformation. These inherent strains can be obtained from the computed results of a TEP FE analysis of the considered basic welded joint. This FE analysis should be previously validated through experiments [9]. This second procedure is preferred and was employed for the evaluation of inherent deformation in this study.

6.2.2.1 TEP FE Analysis and Validation

Based on the geometrical size of the experimental fillet welded joint, an FE analysis model was created with 3D solid elements, as shown in Fig. 6.14. The number of nodes and elements was 22,351 and 18,540, respectively. The welding direction was parallel to the positive direction of the X axis, as shown in Fig. 5. The boundary conditions to fix the rigid body motion are also presented in Fig. 6.14. Temperature-dependent material properties of the specimen (SM490A) are shown in Fig. 6.15 (a: thermal properties; b: mechanical properties). They were used for TEP FE analysis of the examined filet welded joint. Following the experiment, two welding passes were sequentially performed.

Using the in-house JWRIAN (Joining and Welding Research Institute Analysis) coding developed by JWRI, TEP FE computation was conducted to predict welding induced distortion and inherent strain. Figure 6.16 shows the deformed shape after welding and the non-deformed shape of the experimental fillet welded joint. The positions of the measured points were also indicated in Fig. 6.16. The experimental

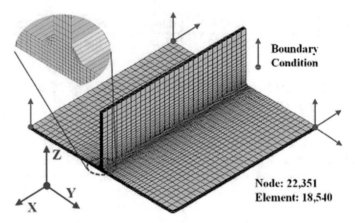

Fig. 6.14 Solid-element model of fillet welded joint

and computed results of welding angular distortion were in good agreement at the measured points, as shown in Fig. 6.17. Because of the temperature gradient through the thickness direction by welding heat input, the angular distortion was produced. Figure 6.17 also shows that the welding angular distortion barely increased along the welding direction.

6.2.2.2   Inherent Strain in Fillet Welded Joint

For the examined fillet welded joint in the actual fabrication process, two welding passes could be carried out by sequential or simultaneous welding processes. Simultaneous welding, in which two welding passes are performed together, can support higher production efficiency than sequential welding. However, it produces much larger longitudinal shrinkage force [10, 11], which is considered the cause of welding induced buckling. Both of these welding processes were examined by FE analysis in this study, and welding induced longitudinal and transverse inherent strains were systematically investigated.

Concerning the sequential welding process in the fillet welded joint, computed results such as transient temperature field, fusion zone shape, welding distortion, and longitudinal and transverse inherent strains produced by welding were directly extracted from previous TEP FE analyses.

Figure 6.18 shows the transient temperature field when the moving heat source with arc torch was passing the selected cross-section of the fillet welded joint. Note that the fusion zone shape can be hardly observed in Fig. 6.18. Almost no heat transfer to the right welded joint occurred when applying the dummy-element technique to the welding sequence, as shown in Fig. 6.18a.

Concerning welding distortion, the deformed shapes and magnitudes of the cross-section of the considered fillet welded joint after the 1st welding pass (left) and 2nd

Fig. 6.15
Temperature-dependent
material properties

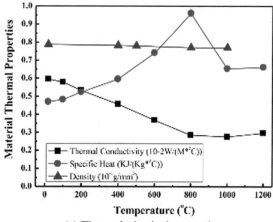

(a) Thermal-physical properties

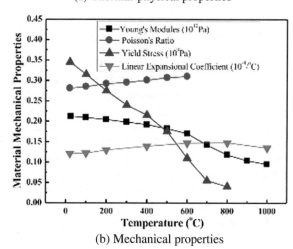

(b) Mechanical properties

Fig. 6.16 Comparison of
non-deformed and deformed
fillet welded joint (scale: 10)

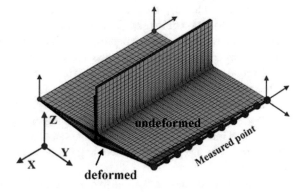

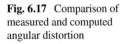

Fig. 6.17 Comparison of measured and computed angular distortion

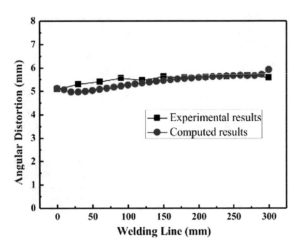

welding pass (right) are shown in Fig. 6.19. This figure also shows that the web rotated to the left side due to the angular distortion produced by the 1st welding pass. It rotated back to its original position but cannot go completely back after the 2nd welding pass. The behavior of web rotational distortion was already reported and discussed with both experimental and computational analyses [12].

Figure 6.20a and b show the contour distributions of longitudinal and transverse plastic strains after the two sequential welding passes finished. Note that the maximum magnitude of either the longitudinal or transverse plastic strains along the welding direction occurred in the area near the end, where the welding was not steady. However, almost uniform distributions along the welding direction occurred in the middle region. Concerning the direction perpendicular to the welding line, note that longitudinal plastic strain took place in a much wider region, and transverse plastic strain was much more concentrated.

Owing to the different generation mechanisms for longitudinal and transverse plastic strains, it is preferred to show the contour distributions of aforementioned plastic strains on a cross-section after the 1st and 2nd welding passes. Figure 6.21 shows the distribution of longitudinal plastic strain. Only the left region of the welded zone presented plastic strain after the 1st welding pass, as shown in Fig. 6.21a. However, when the 2nd welding pass finished, both left and right regions of the welded zone presented plastic strain. Note that the magnitude of the plastic strain in the left region of the welded zone was smaller than just after the 1st welding pass.

The distributions of transverse plastic strain are showed in Fig. 6.22. The plastic strain was also sequentially produced in the regions below the welded zone for the fillet welded joint. Note also that almost the same region and magnitude of transverse plastic strain produced by the 1st welding pass remained after the 2nd welding pass finished.

We selected a perpendicular line to the welding line on the above examined cross-section to examine the welding induced plastic strains closely. Figure 6.23 shows the distributions and magnitudes of longitudinal and transverse plastic strains of nodes

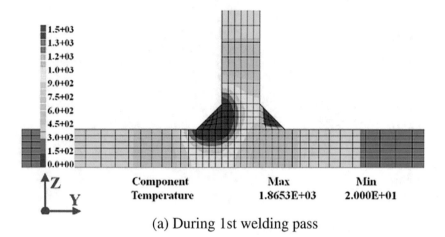

(a) During 1st welding pass

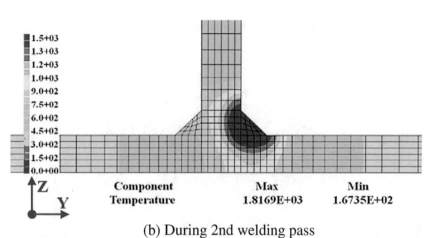

(b) During 2nd welding pass

Fig. 6.18 Transient temperature field when the heat source passed the selected cross-section during sequential welding

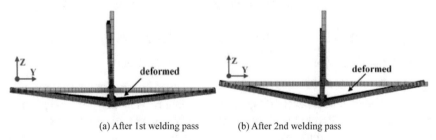

(a) After 1st welding pass   (b) After 2nd welding pass

Fig. 6.19 Front view of welding distortion of fillet welded joint with sequential welding (scale: 10)

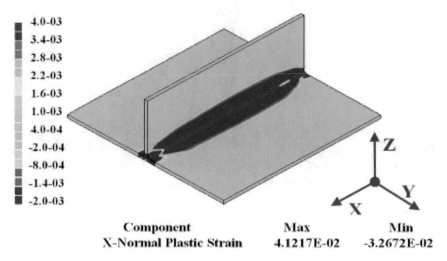

(a) Contour distribution of longitudinal plastic strain

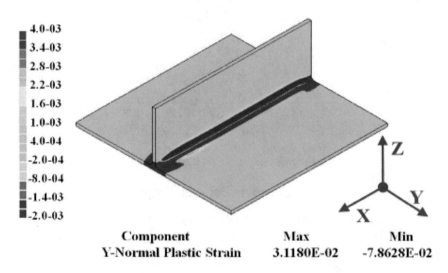

(b) Contour distribution of transverse plastic strain

Fig. 6.20 Contour distribution of plastic strain with sequential welding

on such selected line. After finishing the 1st welding pass, the region of longitudinal plastic strain was slightly wider than that of transverse plastic strain. However, the magnitude of longitudinal plastic strain was much smaller than that of transverse plastic strain. These types of distribution and magnitude were mainly determined by the different self-constraints [13]. Furthermore, Fig. 6.23b shows the significant differences in distributions of longitudinal and transverse plastic strains when the

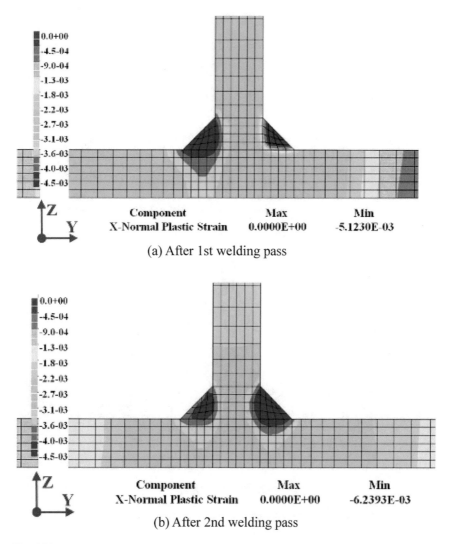

Fig. 6.21 Distribution of longitudinal plastic strain with sequential welding

2nd welding pass finished. Note that there are two evident transverse plastic strain zones and a larger longitudinal plastic strain zone merged by two longitudinal plastic strain zones.

This type of behavior can be clarified by the influence of a subsequent welding pass on an earlier welding pass with self-constraint in longitudinal and transverse directions. Concerning the longitudinal plastic strain, when the 2nd welding pass was carried out, partial plastic strain produced by the 1st welding pass was removed, and new plastic strain occurred and superposed with surplus plastic strain to form the final

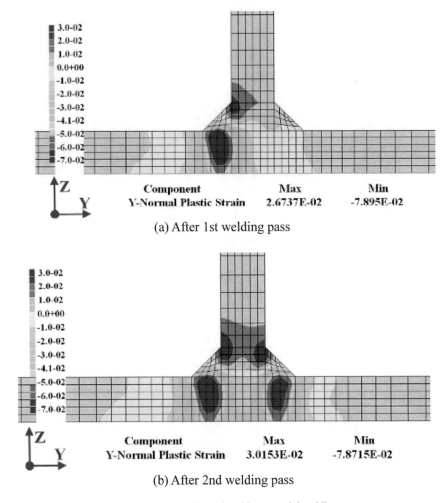

(a) After 1st welding pass

(b) After 2nd welding pass

Fig. 6.22 Distribution of transverse plastic strain with sequential welding

longitudinal plastic strain. However, no transverse plastic strain was removed. Transverse plastic strains produced by two welding passes simply combined to produce the final transverse plastic strain.

In addition, using the above presented FE computation, FE analysis on a fillet welded joint with simultaneous welding was conducted with two welding passes performed together. A much larger fusion zone was obtained in this case, as shown in Fig. 6.24.

Figure 6.25 also shows that obvious transverse shrinkage and symmetrical angular distortion were produced by simultaneous welding. In addition, almost no rotational web distortion took place in comparison with results of sequential welding shown in Fig. 6.19.

Fig. 6.23 Distribution and magnitude of plastic strain with sequential welding

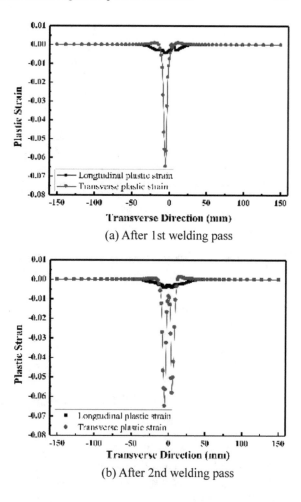

(a) After 1st welding pass

(b) After 2nd welding pass

Figure 6.26 illustrates the almost similar distribution of longitudinal and transverse plastic strains compared with those in sequential welding shown in Fig. 6.20, in which longitudinal plastic strain occurred in a larger region near the welded zone than for transverse plastic strain.

Concerning plastic strains on a cross-section of the examined fillet welded joint assembled by simultaneous welding, a significant difference can be observed in Fig. 6.27 with respect to the computed results shown in Fig. 6.22. Owing to the heating produced by two welding passes together and the fact that no longitudinal plastic strain was removed by subsequent welding, a much larger longitudinal plastic strain region was observed. Note also that completely merged transverse plastic strain was produced by simultaneous welding. The corresponding region was located at the middle, under the two welded zones.

Furthermore, concerning the plastic strains of the points on the line perpendicular to the welding line on the above examined cross-section, Fig. 6.28 indicates that

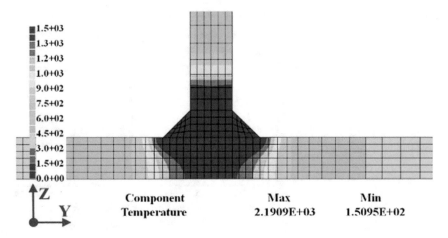

Fig. 6.24 Transient temperature field when the heat source passed the selected cross-section during simultaneous welding

Fig. 6.25 Front view of welding distortion of fillet welded joint with simultaneous welding

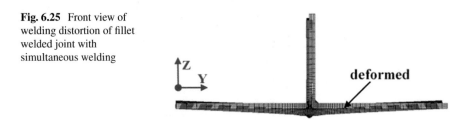

a much larger region of transverse plastic strain was observed in comparison with sequential welding. In addition, a large magnitude of longitudinal plastic strain was generated by simultaneous welding for the same heat input in comparison with sequential welding.

6.2.3 Evaluation of Welding Inherent Deformations

Given that the plastic strains are the dominant components of the inherent strain after welding, the inherent deformation can be evaluated through integration by substituting such strains in the definition given by Eq. (2.8). However, the inherent deformation theory is established by examining the butt weld joint of thin plates. The thickness in Eq. (2.8) denotes the thickness of the welded joint and also the thickness of the plate when the thin plate structure is welded with full penetration. Thus, the inherent deformations for fillet welded joint cannot be directly evaluated with the mentioned approach. Theoretically speaking, based on the physical behavior of

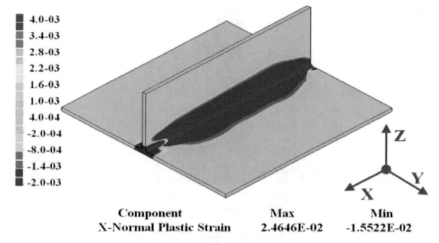

(a) Contour distribution of longitudinal plastic strain

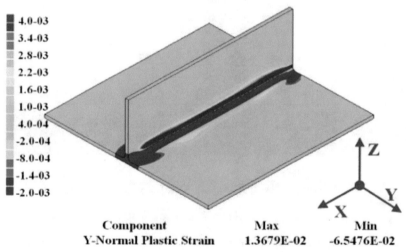

(b) Contour distribution of transverse plastic strain

Fig. 6.26 Contour distribution of plastic strain with simultaneous welding

welding induced inherent deformations, the evaluation process for different components of inherent deformation in fillet welded joints can be achieved with two different
approaches.

In particular, longitudinal inherent shrinkage can be converted to tendon force
for eliminating the influence of the thickness of the welded joint, which is the
consequence of the strong self-constraint in longitudinal direction. This strong self-
constraint implies that the longitudinal inherent shrinkage after welding is extremely
uniform and concentrated closely to the welding line. Therefore, using a force (tendon

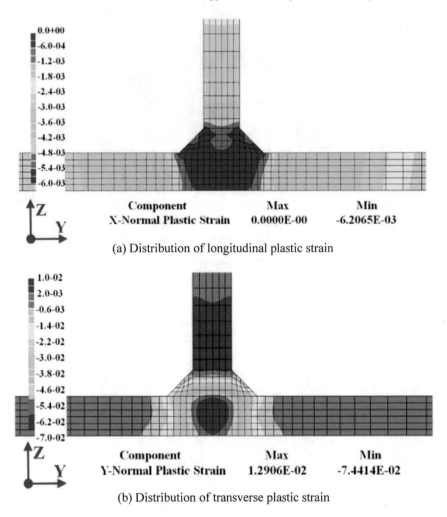

(a) Distribution of longitudinal plastic strain

(b) Distribution of transverse plastic strain

Fig. 6.27 Distribution of plastic strain with simultaneous welding

force) will be better and more rational to represent the actual mechanical behavior of longitudinal welding distortion than applying a uniform shrinkage or deformation. Based on the physical behavior and weak self-constraint for other components of inherent deformation, the transverse inherent shrinkage and bending can be directly evaluated by welding displacement. The longitudinal bending is usually neglected because of its much smaller magnitude.

The target of this study was to predict the welding induced buckling of a skin plate in a ship panel structure, which is almost entirely determined by in-plane inherent deformation. The transverse shrinkage and bending of the web in a fillet welded joint (stiffener) has no influence on that. Thus, in this study, only transverse shrinkage and bending of flange in a fillet welded joint (skin plate) were considered and evaluated.

Fig. 6.28 Distribution and magnitude of plastic strains with simultaneous welding

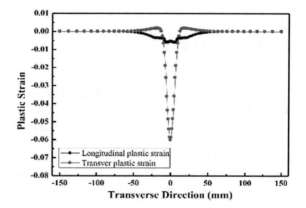

6.2.3.1 Longitudinal Inherent Strain (Tendon Force)

White's parametric study [10] defined the magnitude of tendon force in the following formula based on experimental measurements for a single pass weld:

$$F_{tendon} = 0.2 Q_{net} \tag{6.2}$$

where F_{tendon} is in kN and Q_{net} (J/mm) is the net heat input per unit length.

Later, based on the analysis of heat conduction to calculate the inherent strain in ideal cases and integration of these inherent strains, Wang et al. [11] theoretically proposed the following formula to evaluate the tendon force:

$$F_{tendon} = -0.235 Q_{net} \tag{6.3}$$

For any welded joint, the tendon force can be precisely evaluated by integrating the inherent strain as follows:

$$F_{tendon} = E \times \iint \varepsilon_{longitudinal}^{inherent} dx dy \tag{6.4}$$

Considering the definition of longitudinal inherent deformation, the relationship between tendon force and longitudinal inherent shrinkage is established as follows:

$$F_{tendon} = E \times \iint \varepsilon_{longitudinal}^{inherent} dx dy = E \times h \times \frac{1}{h} \iint \varepsilon_{longitudinal}^{inherent} dx dy \times \delta_{longitudinal}^{inherent} \tag{6.5}$$

where h is the thickness of the welded joint.

Sequential and simultaneous welding for fillet welded joints were examined by the proposed TEP FE analysis and inherent strains were obtained. Then, the tendon force (longitudinal inherent shrinkage force) was evaluated from Eq. (6.4).

The tendon force for each cross-section of the considered fillet welded joint is shown in Figs. 6.29 and 6.30. Figure 6.29 shows the tendon force for sequential welding; in particular, it depicts the magnitude of the tendon force on each cross-section after the 1st and 2nd welding passes. Figure 6.30 shows a comparison of the final tendon force on each cross-section for sequential and simultaneous welding. After two welding passes with different welding processes, sequential welding produced a much smaller tendon force because the partial tendon force of the 1st welding pass was removed.

According to the almost uniform distribution in the center region shown in Figs. 6.29 and 6.30 and ignoring the end effect, Table 6.5 gives the constant magnitude of tendon force for different cases.

Fig. 6.29 Comparison of tendon force after 1st and 2nd welding passes with sequential welding

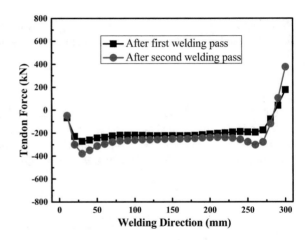

Fig. 6.30 Comparison of tendon force for sequential and simultaneous welding

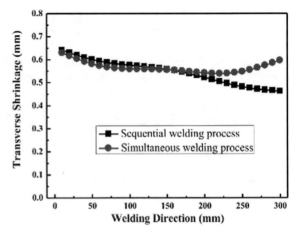

Table 6.5 Evaluated tendon force for different welding processes

	Sequential welding		Simultaneous welding
	After 1st welding	After 2nd welding	
	– 215.877 kN	– 259.055 kN	– 481.425 kN

Concerning the tendon force after the 1st welding pass in the case of sequential welding, its magnitude was calculated using Eq. (6.6) as follows:

$$
\begin{aligned}
F_{tendon} &= -0.235 Q_{net} = -0.235 \times \frac{\eta \times U \times I}{v} \\
&= -0.235 \times \frac{0.74 \times 28\text{V} \times 185\text{A}}{4.2 \text{ mm/s}} = -214.476\text{kN}
\end{aligned}
\tag{6.6}
$$

There was a good agreement for the magnitude of the tendon force with one welding pass between the theoretical formula and TEP FE analysis. When sequential welding was considered, a remained factor (k) had to be used. The remained factor of the current welding pass was set to 1 because there was no subsequent welding to be performed and all the tendon force produced by the current welding pass remained. By contrast, the remained factor of the previous welding pass changed from 0 to 1, depending on the distance of the current and subsequent welding passes. When the current welding pass was far away, the tendon force produced by the previous welding pass totally remained; the tendon force was completely removed by the current welding pass when the two passes were overlapped and the remained factor of the previous welding pass was fixed to 0.

For a fillet welded joint, the following calculation provides the tendon force after the 2nd welding pass:

$$
\begin{aligned}
F_{tendon} &= \sum k_i \times (-0.235 Q_{net}) = k_i \times (-0.235 Q_{net}) + k_2 \times (-0.235 Q_{net}) \\
&= 0.2 \times (-0.235 Q_{net}) + 1 \times (-0.235 Q_{net}) = (-42.895\text{kN}) + (-214.476\text{kN}) \\
&= -257.371\text{kN}
\end{aligned}
\tag{6.7}
$$

6.2.3.2 Transverse Inherent Shrinkage and Bending

Following the above discussion, transverse inherent shrinkage and bending for different welding processes were directly estimated from computed results of TEP FE analysis.

Figure 6.31 shows the inherent transverse shrinkage on each cross-section in the cases of sequential and simultaneous welding. The same distribution and magnitude were basically observed. However, a significant difference of transverse bending between sequential and simultaneous welding was found, as shown in Fig. 6.32. This

Fig. 6.31 Comparison of
transverse shrinkage in the
cases of sequential and
simultaneous welding

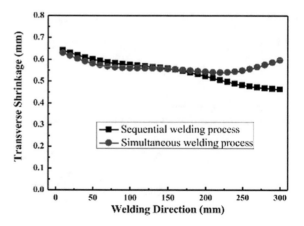

Fig. 6.32 Comparison of
angular distortion in the
cases of sequential and
simultaneous welding

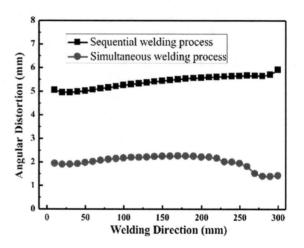

can be clarified by the fact that there was a small temperature gradient through the
thickness of the plate when simultaneous welding was performed, thereby producing
a smaller angular distortion.

Ignoring the end effect, the constant magnitudes of inherent transverse shrinkage
and bending are summarized in Table 6.6.

According to the magnitude of inherent deformation in Tables 6.5 and 6.6 for
different welding processes, simultaneous welding with two welding passes to

Table 6.6 Evaluated
transverse shrinkage and
bending for different welding
process

Sequential welding		Simultaneous welding	
Transverse shrinkage (mm)	Transverse bending (rad)	Transverse shrinkage (mm)	Transverse bending (rad)
0.54829	0.036	0.553788	0.01429

improve the production efficiency gives rise to a much larger longitudinal inherent deformation (tendon force) but much smaller angular distortion than those generated in sequential welding. It can be theoretically clarified as follows: there should be no 2nd welding pass to remove the tendon force produced in the previous step, and small temperature gradient in thickness generates a small angular distortion.

6.2.4 Elastic FE Analysis with Inherent Deformations

When the inherent deformations of a basic welded joint included in the considered actual welded structure are evaluated, they can be employed for predicting welding distortion of the actual welded structure through elastic FE analysis. In this study, elastic FE analysis with inherent deformation was an ideal and practical method to predict welding distortion of the considered ship panel welded structure.

6.2.4.1 Welding Induced Buckling in Ship Panel Structure

Figure 6.33 shows the considered ship panel welded structure, which was assembled from a skin plate (16,400 mm × 4,100 mm), longitudinal stiffeners with L section (angle-bar), and transverse stiffeners with T section (fillet-bar). The longitudinal and transverse heights of the stiffeners were 100 and 368 mm, respectively. The thickness of all stiffeners was 6 mm. The shell-element model of this considered ship panel is shown in Fig. 6.34. The boundary conditions and locations of the examined points on lines 1 and 2 are also shown in Fig. 6.34.

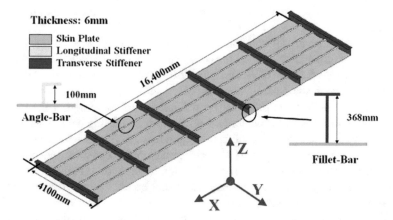

Fig. 6.33 Overview of the considered ship panel

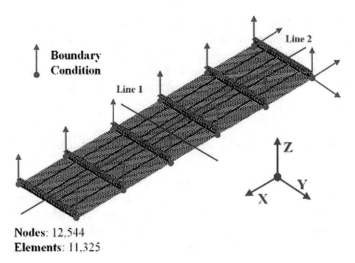

Fig. 6.34 Ship panel structure model with shell elements and constraint

The assembling sequence for this ship panel welded structure was joining the skin plate and longitudinal stiffeners first, and then transverse stiffeners were welded sequentially.

During the assembly process to join the longitudinal and transverse stiffeners to the skin plate, different welding processes were employed. The skin plate and longitudinal stiffeners were welded by simultaneous welding with an automatic welding machine to improve the production efficiency. The skin plate and transverse stiffeners were also welded by sequential welding, which was performed through one-by-one welding pass. By applying the different inherent deformations for the corresponding welding process, the welding induced buckling could be predicted through elastic FE analysis.

Using the inherent deformations evaluated by TEP FE analysis in previous sections, the welding distortions of the considered ship panel structure was obtained. It is shown in Figs. 6.35 and 6.36. These distortions were predicted by elastic FE analysis employing the small deformation and large deformation theorems. Figure 6.35 shows that the deflections produced by welding clearly appeared at the two side edges. The deflection in the center part of the considered ship panel structure was relatively small. This type of welding distortion resulted from the fact that the parts located at the two side edges were almost free and underwent no strong constraint. In the case of small deformation theory, the angular distortion (transverse bending) produced by welding was the primary cause that generated this form of welding distortion during the assembly process.

However, in the case of large deformation theory, the computed results showed that the welding induced buckling was produced not only at the two side edges, but also at the central region of the ship panel, as shown in Fig. 6.36. This deflection

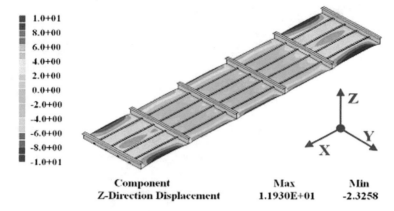

Component	Max	Min
Z-Direction Displacement	1.1930E+01	-2.3258

Fig. 6.35 Welding distortion of ship panel employing small deformation theory (scale: 10, unit: mm)

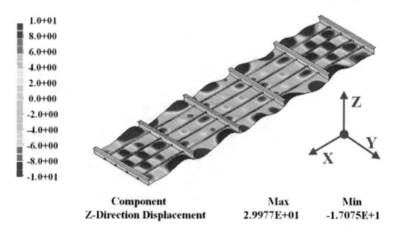

Component	Max	Min
Z-Direction Displacement	2.9977E+01	-1.7075E+1

Fig. 6.36 Welding distortion of ship panel employing large deformation theory (scale: 10, unit: mm)

is the buckling deformation, which was mainly determined by the in-plane inherent deformations produced by welding during the assembly process [11].

To examine the difference between the computed results employing the small and large deformation theorems closely, the computed deflections of points on lines 1 and 2 defined in Fig. 6.34 are plotted in Figs. 6.37 and 6.38, respectively. Note that no deflections occurred at the positions of the stiffeners during the welding processes. Figures 6.37 and 6.38 show that between the cases considering small and large deformations, the distribution of deflection (deformed mode) and magnitude of deflection presented significant differences in the points on lines 1 and 2. These differences are summarized next:

Fig. 6.37 Deflection of points on line 1 in transverse direction

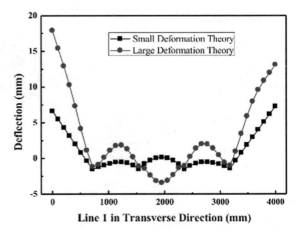

Line 1 in Transverse Direction (mm)

Fig. 6.38 Deflection of points on line 2 in longitudinal direction

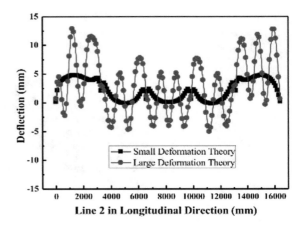

Line 2 in Longitudinal Direction (mm)

(1) Figures 6.37 and 6.38 show that the case of small deformation exhibits deformation with hungry horse mode. They are all one half-wave bending deformations in the upward direction because of angular distortion. However, the deflection mode was different for the case of large deformation. The plates of the ship panel structure were deformed with multi half-wave mode. This type of tendency is evident in the side edge points shown in Fig. 6.37 compared with that at the center, as shown in Fig. 6.38.

(2) The magnitude of deflection is another difference for the two computed cases. For the case of the large deformation, the computed deflection was always greater than for small deformation, as shown in Figs. 6.37 and 6.38.

Overall, for the case of small deformation theory, the welding distortion of the considered ship panel structure was mainly produced by angular distortion. Moreover, the welding induced buckling generated by in-plane inherent deformation in the case of large deformation theory was a consequence of non-linear behavior.

6.2.4.2 Buckling Prediction of Stiffened Thin Plate Sections

The above validation of elastic FE analysis enables the prediction of welding distortion in fabrication of lightweight ship panels as an actual application of elastic FE analysis. In particular, a unit ship panel with 6-mm thickness plates was selected as case study. It was assembled from a skin plate with 13,120 mm in length and 3,205 mm in width, 3 longitudinal stiffeners with 100 mm in height as an L section (angle bar), and 4 transverse stiffeners with 394 mm in height as a T section (fillet bar).

The shell-element model of the considered unit ship panel is depicted in Fig. 6.39, in which different parts to be fitted and then welded together are indicated by different colors. Between the adjacent parts, previously evaluated inherent deformation was applied to represent the mechanical features of the welding line. As shown in Fig. 6.39, the boundary condition was employed to fix the rigid body motion. Lines 1 and 2 were also selected to investigate the out-of-plane welding distortion closely. Applying inherent deformation and the large deformation theory, elastic FE analysis of the shell-element model was carried out, and out-of-plane welding distortion was predicted. In particular, the potential buckling behavior generated by in-plane welding shrinkage was observed in the fabrication of the examined unit ship panel.

Significant features of buckling from not only the magnitude but also the deformed mode of out-of-plane welding distortion were observed, as shown in Fig. 6.40. Therefore, it can be concluded that the inherent deformations generated by current welding procedures can produce the examined ship panel buckle. Figure 6.41 shows the computed out-of-plane welding distortions of points on lines 1 and 2. Note that welding induced buckling with half-wave deformed mode occurred.

Out-of-plane welding distortion was produced by three causes: welding bending moment, in-plane inherent shrinkage, and bending moment due to shrinkage force.

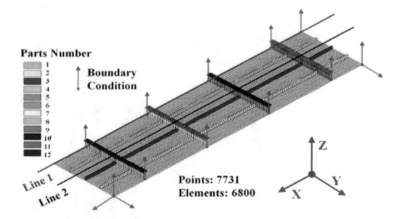

Fig. 6.39 Shell-element model of studied unit ship panel and computational boundary condition (lines 1 and 2)

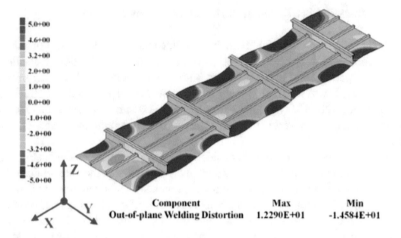

Fig. 6.40 Out-of-plane welding distortion from elastic FE analysis only employing in-plane inherent shrinkage

Fig. 6.41 Out-of-plane welding distortions of points on selected lines

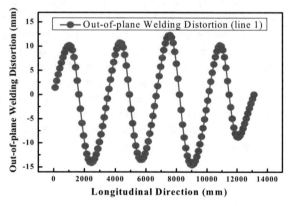

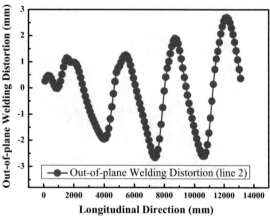

When welding induced buckling occurred, buckling deformation was the dominant component that contributed the out-of-plane welding distortion. Likewise, in-plane inherent shrinkage was the dominant cause of out-of-plane welding distortion. Other components of inherent deformation and initial deflection were jointly considered as the disturbance that triggered the welding induced buckling. They did not determine the buckling mode but influenced the magnitude of out-of-plane welding distortion. Note that the relatively large bending deformation produced by the welding bending moment may obscure the phenomenon of welding induced buckling.

6.2.5 Techniques for Welding Buckling Prevention

Following the above discussion, the influence of welding procedure patterns on welding induced buckling was examined. In particular, the magnitudes of inherent deformation produced by the proposed welding procedure patterns were evaluated in advance. Then, these inherent deformations were applied for elastic FE analysis to predict out-of-plane welding distortion of the examined ship panel.

In actual welding for assembling many parts of the considered ship structure, different welding procedure patterns such as zigzag welding and intermittent welding were employed to replace parallel continuous welding and reduce the welding distortion and fabrication cost. Figure 6.42 illustrates the commonly used welding procedure patterns for assembling stiffened ship panels. Other three welding procedure patterns were considered to fabricate the examined ship panel through computational investigation. These welding distortions were compared with that of parallel continuous welding.

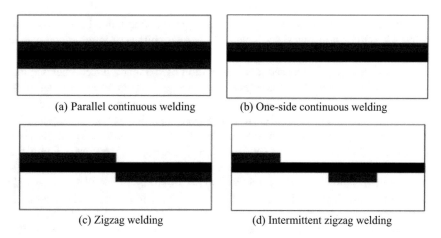

(a) Parallel continuous welding   (b) One-side continuous welding

(c) Zigzag welding   (d) Intermittent zigzag welding

Fig. 6.42 Illustration of commonly used welding procedure patterns

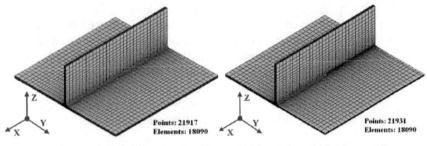

(a) FE analysis model for one-side continuous welding   (b) FE analysis model for zigzag welding

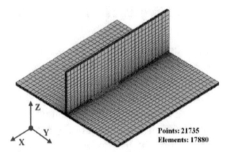

(c) FE analysis model for intermittent zigzag welding

Fig. 6.43 Solid-element models for different welding procedure patterns

Figure 6.43 shows the thermal elastic-plastic FE model with solid elements for the three examined welding procedure patterns: (a) one-side continuous welding; (b) zigzag welding, and (c) intermittent zigzag welding. Using the aforementioned in-house code again for thermal and mechanical analyses with FE computations, the distribution and magnitude of plastic strain in the longitudinal direction were computed, as shown in Fig. 6.44. Note that the distributions of longitudinal plastic strain presented significant differences.

As mentioned in previous sections, prevention of welding induced buckling can be achieved by advanced welding procedures to generate less inherent deformation. In actual cases, intermittent zigzag welding is often employed because of its practicability and flexibility. This allows to replace parallel continuous welding for reducing welding buckling distortion in the fabrication of unit ship panels [14]. Figure 6.42c illustrates the proposed procedure to accomplish the fillet welding; the welding conditions are the same as in the experimental case summarized in Table 6.4. The computed angular distortion and welding plastic strains of the fillet welded joint with intermittent zigzag welding procedure were obtained through non-linear transient TEP FE analysis. Distribution and magnitudes of the longitudinal plastic strain are shown in Fig. 6.44c. They were employed to evaluate the tendon force. The computed longitudinal and transverse plastic strains generated by sequential and zigzag welding procedures on the middle cross-section were compared, as shown in Fig. 6.45. Other components of inherent deformation were also evaluated; they are

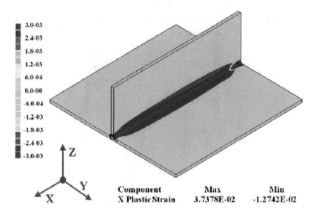

(a) Longitudinal plastic strain produced by one-side continuous welding

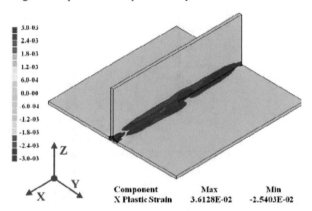

(b) Longitudinal plastic strain produced by zigzag welding

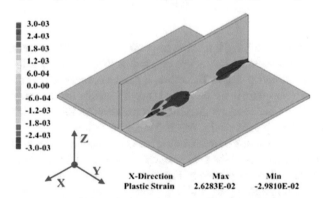

(c) Longitudinal plastic strain produced by intermittent zigzag welding

Fig. 6.44 Computed longitudinal plastic strain with different welding procedures

Fig. 6.45 Comparison of longitudinal and transverse plastic strains generated by sequential and zigzag welding procedures

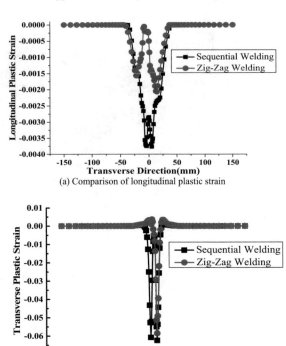

(a) Comparison of longitudinal plastic strain

(b) Comparison of transverse plastic strain

Table 6.7 In-plane inherent deformation generated by intermittent zigzag welding procedure

Tendon force	Transverse inherent shrinkage
−124.655 (kN)	0.10779 (mm)

summarized in Table 6.7. Note that the in-plane welding inherent deformation with intermittent zigzag welding procedure was below the previously calculated critical values of inherent deformation. Therefore, it can be concluded that the intermittent zigzag welding procedure in the examined case effectively decreased the magnitudes of inherent deformation and avoided the occurrence of welding induced buckling.

The longitudinal plastic strains on each cross-section normal to the welding line were all integrated with the corresponding areas to calculate inherent deformations. The resulting magnitudes of the tendon force (longitudinal inherent shrinkage force) along the welding line are plotted in Fig. 6.46. This figure also shows that the side continuous welding and zigzag welding presented almost the same magnitude of tendon force along the welding line. This behavior was due to the fact that the same heat input per unit welding length was applied to the welding line by the welding arc. In particular, side continuous welding was performed in the side of the fillet welded joint. Two sides of the fillet welded joint were exclusively and sequentially heated

Fig. 6.46 Comparison of tendon force produced by different welding procedures

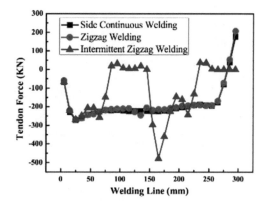

by the welding arc under zigzag welding. The tendon force caused by intermittent zigzag welding underwent a drastic change along the welding line and gave rise to an evident different distribution compared with that in the cases of the other two examined welding procedure patterns.

To compare the influence of welding procedure patterns on welding induced buckling in the fabrication of the examined ship panel, the magnitudes of in-plane inherent shrinkages are summarized in Table 6.8. The following conclusions can be drawn:

(1) In-plane inherent shrinkages produced by one-side continuous welding and zigzag welding basically coincided because the same heat input per unit welding length was applied.

(2) When parallel continuous welding was replaced by one-side continuous welding (zigzag welding), the tendon force (longitudinal inherent shrinkage force) was slightly reduced. However, transverse inherent was clearly reduced, with approximately 50% of that shrinkage caused by parallel continuous welding. This behavior can be explained by the fact that the second welding pass in the fillet welded joint reduced the partial longitudinal plastic strain generated by the first welding pass and had no influence on the transverse plastic strain previously generated [15].

Table 6.8 Magnitudes of in-plane inherent shrinkages produced by different welding procedure patterns

	Tendon force (kN)	Transverse inherent shrinkage (mm)
Parallel continuous welding	−259.055	0.54829
One-side continuous welding	−214.564	0.29555
Zigzag welding	−211.582	0.25676
Intermittent zigzag welding	−124.655	0.10779

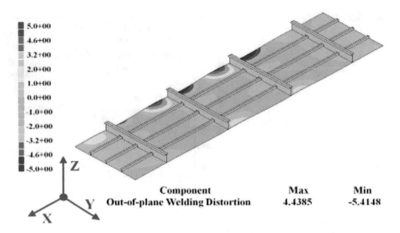

Fig. 6.47 Out-of-plane welding distortion from elastic FE analysis by employing intermittent zigzag welding procedure (scale: 10)

(3)   Intermittent zigzag welding can significantly reduce the in-plane inherent shrinkages because of the less heat input per unit welding length.

To examine the out-of-plane welding distortion closely, elastic FE analysis with inherent deformation generated by intermittent zigzag welding procedure was carried out again to predict the out-of-plane welding distortion of the examined unit ship panel. The mitigation effort with intermittent zigzag welding procedure to reduce the out-of-plane welding distortion is illustrated in Fig. 6.47. Note that not only the magnitude but also the deformed mode significantly changed. Out-of-plane welding distortions of points on lines 1 and 2 shown in Fig. 6.39 were also plotted and compared with data, as summarized in Fig. 6.48. This figure clearly shows that the intermittent zigzag welding procedure produced negligible out-of-plane welding distortion at the middle region of the examined unit ship panel and acceptable out-of-plane welding distortion at the edge region because the edge region is usually corrected after welding to be assembled with other components or blocks in actual fabrication. Intermittent zigzag welding procedure precluded welding induced buckling because of the reduced in-plane inherent deformation. However, the out-of-plane welding distortion could not be prevented because of the contribution of the bending moment due to the shrinkage force.

Mitigation of the out-of-plane welding distortion caused by welding induced buckling at the middle region with different welding procedure patterns was the main objective of this study. Therefore, fabrication of ship panels with an appropriate intermittent zigzag welding procedure that can produce the welded joint with suitable mechanical features is an efficient and practical approach for controlling the welding induced buckling and manufacturing thin-plate welded structures with high dimensional quality.

Fig. 6.48 Comparison of out-of-plane welding distortion with different welding procedure patterns

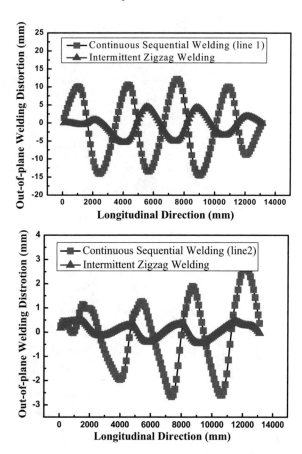

6.3   Application of Accurate Fabrication for Container Ship

With the fast increase of international trade, container vessels, in particular ultra-large container ships, are becoming increasingly demanded owing to advantages such as efficiency, convenience, and safety. An ultra-large container ship with 20,000TEU is shown in Fig. 6.49. It is becoming the dominant type in contrast with ships with lower cargo capacity.

Generally, an ultra-large container ship is assembled with lots of panel blocks and curved sections and two major structures: a watertight transverse bulkhead structure and a torsion box structure, as shown in Fig. 6.50. Owing to mechanical performance considerations and TEU loading/unloading efficiency, panel blocks with the watertight transverse bulkhead and torsion box structures required accurate fabrication with much less out-of-plane welding distortion. Moreover, different plate thicknesses and materials, in addition to partially penetrated welded joints, were employed to reduce the fabrication cost and schedule. The application of these aspects during the actual fabrication of the examined panel blocks made our investigation more complex

Fig. 6.49 Overview of ultra-large container ship with 20000TEU

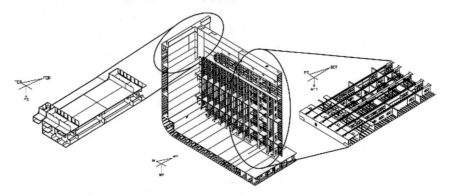

Fig. 6.50 Considered welded structure with watertight transverse bulkhead structure (right) and torsion box structure (left)

in terms of accurate prediction and mitigation of out-of-plane welding distortion. Therefore, both advanced and efficient computational approaches for welding distortion prediction and mitigation during the accurate fabrication of large complex ship panel blocks are indeed desired in modern shipbuilding. In addition, for welding distortion investigation during the fabrication of ultra-large container ships, it is better to predict the distribution and magnitude of out-of-plane welding distortion with an advanced computational approach, and then consider the generation mechanism and propose advanced techniques for welding distortion correction in the actual shipbuilding.

As a result of the rapid development of computational approaches, in particular the finite element method, thermal elastic-plastic FE computation was widely accepted and employed to solve the welding distortion problem, but only for welded joints and simple welded structures owing to its large consumption of computer memory and long computing time. Elastic FE analysis with shell-element model is an ideal and practical computational approach for prediction and mitigation of welding distortion during the fabrication of ship structures.

Two major and essential structures as watertight transverse bulkhead and torsion box in ultra-large containers with 20000TEU were researched. Welding procedures with different welding methods and welding conditions were carried out for the considered welded structure fabrication. Measurements of out-of-plane welding distortion were also experimentally conducted with Total Station. Typical welded joints with different materials and joining forms were classified and summarized. Welding-induced inherent deformations were evaluated based on efficient TEP FE computations with ISM and OpenMP parallel computation for all typical welded joints, while the integration method of welding inherent strain and inverse analysis of displacement were both employed and compared for accuracy validation and inherent deformation evaluation. A database of welding inherent deformation for ultra-large container ships was then created, and empirical formulae relating welding inherent deformations to welding heat input for different welded joints were proposed through a linear regression approach for subsequent elastic FE computation. Applications of welding distortion prediction from elastic FE computation with shell-element model and welding inherent deformation as mechanical loading were practiced for the considered welded structures. Not only the deformed shape but also the magnitudes of out-of-plane welding distortions measured were in good agreement with the computed results. The influence of welding sequence on welding distortion mitigation was studied during the fabrication of a watertight transverse bulkhead structure, and design optimization of a thick-plate butt welded joint with X type groove was examined for welding distortion mitigation during the fabrication of a torsion box structure. Reduction effects of out-of-plane welding distortion with different mitigation practices were observed and demonstrated with the employed elastic FE computation. Precision fabrication of ultra-large containers can be achieved with the computed results and corresponding analysis.

6.3.1 Examined Structures and Welding Experiments

The watertight transverse bulkhead structure is important for TEU delivery and fixation. It is usually assembled from an angle steel, a stiffener, frame girders, and slideway structures. The examined structure shown in Fig. 6.51 is a typical watertight transverse bulkhead structure in ultra-large container ships with 20,000TEU, which are ships with 29.79 m in length, 16.50 m in width, and 1.95 m in height. In particular, the examined structure is a two-layer stiffened structure. The bulkhead structure was joined by 11 rectangle steel plates; 38 angle steel structures and 17 frame girders were then welded to the bulkhead structure with fillet welding along the length direction; 24 stiffeners were finally welded to the bulkhead structure with fillet welding along the width direction. Additionally, some stiffeners and frame girders were designed with a hole for piping. Owing to the complicated welding procedure of different thickness plates, welding distortion was clearly generated. This influenced

Fig. 6.51 Watertight
transverse bulkhead structure

the dimensional precision and subsequent fitting procedure with other blocks. From the taken measurements, out-of-plane welding distortion reached a large magnitude of approximately 20 mm.

Concerning the torsion box structure shown in Fig. 6.52, it is a two-layer structure featuring 18.50 m in length, 7.31 m in width, and 3.50 m in height. A great deal of longitudinal and transverse stiffeners act as supporting components at the center region and thick plates with thickness in the range 60–85 mm were applied for the hatch coaming structure. In addition, two skin plates acting as outside and inside bulkheads were assembled from four rectangular plates with different thicknesses. Six longitudinal stiffeners with continuous geometry were welded to the skin plates. In transverse direction, five frame girders were placed, all of them perforated to ensure continuity of longitudinal stiffeners and reduce the weight with a rectangular form. In addition, a deck was fabricated from three thick plates with 85-mm thickness. They were joined to the skin plates by four longitudinal and continuous large-size stiffeners.

Based on a diversity of welding conditions, materials, plate thicknesses, and welded joint forms, Tables 6.9 and 6.10 summarize the typical butt and fillet welded joints in both watertight transverse bulkhead and torsion box structures of the examined ultra-large container ship with 20000TEU. Note that FYS stands for Y-type groove with FCB (flux copper backing) welding; AXS stands for X-type groove with double-side GMAW (gas metal arc welding) or SAW (submerged arc welding);

Fig. 6.52 Torsion box
structure

Table 6.9 Geometrical features and materials of typical butt welded joints

Typical welded joint	Groove design	Thickness (mm)	Materials
FYS-12-12		12–12	AH32
FYS-12-16		12–16	
FYS-16-16		16–16	
COVN-10-10		10–10	AH32
AI-10-14		10–14	AH32
AI-14-22		14–22	AH32-AH36
AXS-85-85		85–85	EH40
AXS-60-60		60–60	DH32
AXS-85-60		85–60	EH40-DH32
COXS-85-85		85–85	AH36
COXS-40-40		40–40	AH32
COVN/3-14-14		14–14	AH32
AYN-15-18		15–18	AH32

Table 6.10 Geometrical features and materials of typical fillet welded joints

Typical welded joint	Groove design	Thickness (mm)	Materials
FP-9-14		9–14	AH32
FP-10-13		10–13	
FP-13-16		13–16	
FP-22-16		22–16	AH32-AH36
PP-80-80-15		80–80	EH47
PP-80-80-20		80–80	EH47
PP-80-80-27		80–80	EH47
PP-85-85-33		85–85	EH40

COVN stands for V-type groove with GMAW; and AI stands for I-type groove with double-side SAW. For fillet welded joints, FP denotes a welded joint with full penetration whereas PP denotes partial penetration of web plate. The welding conditions of FCAW (flux cored arc welding) and SAW during ship structure fabrication are listed in Tables 6.11 and 6.12, respectively. Different welding processes such as root welding, fill welding, and cap welding were individually carried out with their corresponding welding conditions.

During the fabrication of the watertight transverse bulkhead and torsion box structures, out-of-plane welding distortion was measured in the shipyard with Total Station, as shown in Fig. 6.53. These measurements were later applied to validate the accuracy of the computed results.

Table 6.11 Welding conditions of FCAW with CO_2

	Current (A)	Voltage (V)	Velocity (mm/min)	Heat input (kJ/mm)
Root	185–210	24–27	95–120	2.21–3.69
Fill	230–260	25–29	300–375	0.92–1.53
Cap	220–250	25–29	305–380	0.86–1.44

Table 6.12 Welding conditions of SAW

	Current (A)	Voltage (V)	Velocity (mm/min)	Heat input (kJ/mm)
Root	520–590	25–29	340–420	1.84–3.07
Fill	660–750	28–35	440–510	2.17–3.61
Cap	650–730	29–34	400–500	2.25–3.75

Fig. 6.53 Welding distortion measurements with Total Station

6.3.2   Creation of Inherent Deformation Database

Taking into account the large and complex target welded structure, it was a challenge to predict the welding distortion with conventional thermal elastic-plastic FE computation because of the large consumption of computer resources and long computing time [1]. Elastic FE computation with welding inherent deformation is an ideal and practical numerical approach for welding mechanical response investigation. The magnitudes of welding inherent deformation are desired in advance for elastic FE analysis. Generally, two methods for inherent deformation evaluation are employed: inverse analysis with measurement of welding distortion [16], and integration with computed inherent strain, which is based on the definition of inherent deformation [1, 17].

Based on Tables 6.9 and 6.10, typical welded joints with different joint formations, materials, and plate thicknesses were created with a solid-element model. Then, fast thermal elastic-plastic FE computation was employed to investigate the thermal and mechanical responses during the multi-pass welding procedure for each typical welded joint. Once the computed results were obtained, welding inherent deformations were evaluated first by the inverse method with computed welding distortion, and second by the integration method with computed plastic strain. The accuracy was also validated by comparing the computed results from elastic FE analysis with welding inherent deformation as input loading with thermal elastic-plastic FE analysis.

6.3.2.1   Solid-Element Model of Welded Joints

As mentioned above, welding inherent deformation can be employed for welding distortion prediction of large ship structures with elastic FE computation. Thus, welding inherent deformation must be previously evaluated. The integration method with computed inherent strain was first applied. Thermal elastic-plastic FE computation for typical welded joints was carried out later.

To clarify the difference of typical welded joints, six welded joints with solid-element model were introduced and considered for evaluation of welding inherent deformation:

(1)   Figure 6.54 shows welded joint FYS-16-16, which is a Y-type groove with FCB welding and plate thicknesses of 16 mm;

(2)   Figure 6.55 shows AI-14-22, which is an I-type groove with double-side SAW and different plate thicknesses of 14 mm and 22 mm, respectively;

(3)   Figure 6.56 shows AXS-85-85, which is an X-type groove with double-side GMAW or SAW and plate thicknesses of 85 mm;

(4)   Figure 6.57 shows COVN-14-14, which is a V-type groove with GMAW and plate thicknesses of 14 mm;

(5)   Figure 6.58 shows FP-22-16, which is a full-penetration joint with single-side GMAW, web thicknesses of 22 mm and flange thickness of 16 mm;

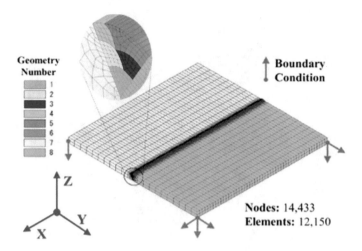

Fig. 6.54 Solid-element model and arrangement of welding passes for FYS-16-16

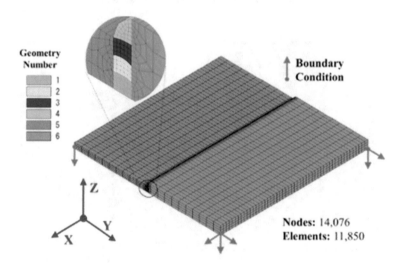

Fig. 6.55 Solid-element model and arrangement of welding passes for AI-14-22

(6) Figure 6.59 shows C-85-85, which is a partial-penetration joint with double-side GMAW and flange and web thicknesses of 85 mm.

6.3.2.2 Fast TEP FE Computation

The in-house code was programmed and implemented to support the theorem of heating conduction, elastic-plastic mechanics, and the finite element method. We used such code to carry out thermal analysis to represent the transient temperature

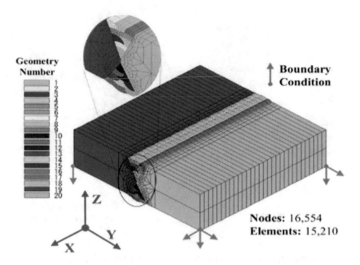

Fig. 6.56 Solid-element model and arrangement of welding passes for AXS-85-85

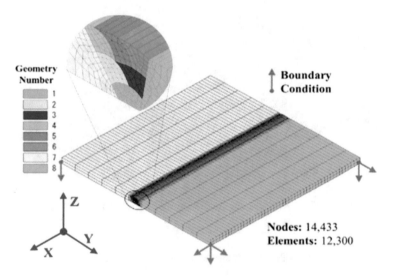

Fig. 6.57 Solid-element model and arrangement of welding passes for COVN-14-14

distribution during the welding process and predict the shape of the welded pool for
the six aforementioned welded joints illustrated from Figs. 6.54, 6.55, 6.56, 6.57,
6.58 and 6.59. In addition, dummy element technology was employed to consider
the non-welded pass in the multi-pass welding procedure. A body heat source with
uniform power density Q (W/m^3: welding arc energy/volume) was applied to model
the welding arc. Moreover, heat losses and the surface heat flux due to thermal

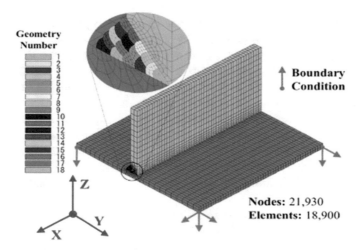

Fig. 6.58 Solid-element model and arrangement of welding passes for FP-22-16

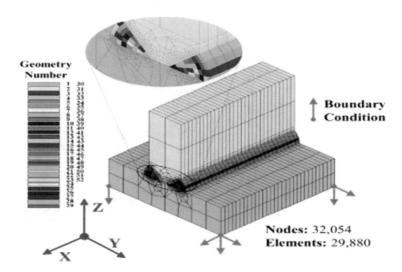

Fig. 6.59 Solid-element model and arrangement of welding passes for C-85-85

convection and radiation were also taken into account with combined convection and radiation boundary flux q (W/m^2) on all external surfaces expressed as follows:

$$q = q_c + q_r$$
$$= h(T - T_{room}) + \varepsilon C[(T + 273)^4 - (T_{room} + 273)^4]$$
$$= [h + \varepsilon C[(T + 273)^2 + (T_{room} + 273)^2][(T + 273) + (T_{room} + 273)]](T - T_{room})$$
$$= \mathrm{h}(T)(T - T_{room}) \tag{6.8}$$

where q_r and q_c denote the heat convection and radiation, respectively; h is the coefficient of heat convection; ε is the emissivity of the objective, which is equal to one for an ideal radiator; C is the Stefan-Boltzmann constant; $h(T)$ denotes the combined convection and radiation boundary flux coefficient, whose value was fixed to 5.7 $\times$ 10^{-6}w/(°C*m²) regardless of the temperature variation during FE computation; finally, $(T - T_{room})$ denotes the increased temperature magnitude.

Then, transient temperature distributions were considered as input thermal load to investigate the mechanical response in terms of welding distortion and plastic strain, among other phenomena. Note that the rigid body motion was fixed to be the mechanical boundary conditions illustrated from Figs. 6.54, 6.55, 6.56, 6.57, 6.58 and 6.59. Figure 6.60 shows the temperature-dependent thermal and mechanical material properties of AH32.

Figure 6.61 shows the computed transient temperature distribution of cap welding and shape profile of the welded pool. Note that the region with temperature above the steel molten point was close to the welded groove. Figure 6.62 illustrates the predicted out-of-plane welding distortion after welding the FYS-16-16 welded joint. Note that the deformed rate was 5. The deformed shape was amplified 5 times for clear

Fig. 6.60
Temperature-dependent
material properties of AH32

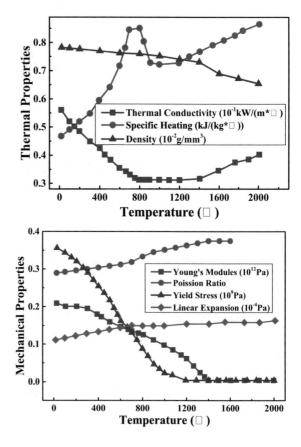

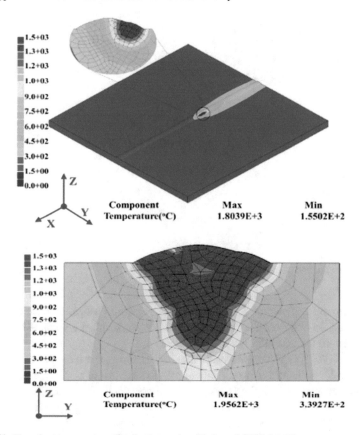

Fig. 6.61 Transient temperature distribution and welded pool (FYS-16-16)

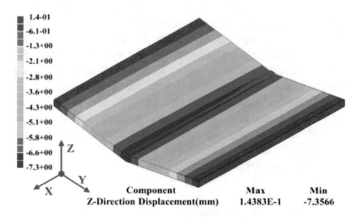

Fig. 6.62 Out-of-plane welding distortion of FYS-16-16 (deformed rate: 5)

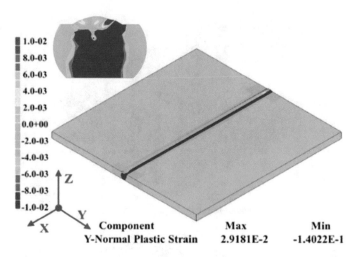

Fig. 6.63 Distribution of transverse plastic strain of FYS-16-16

observation and analysis of deformed tendency during post-processing of welding computation. In addition, the distribution of transverse plastic strain after welding the FYS-16-16 welded joint is shown in Fig. 6.63.

For the other typical welded joints mentioned above, the temperature field, out-of-plane welding distortion, and residual plastic strain were also examined through a thermal-mechanical-coupling computational approach. The computed results are shown in Fig. 6.64 for AI-14-22, Fig. 6.65 for AXS-85-85, Fig. 6.66 for COVN-14-14, Fig. 6.67 for FP-22-16, and Fig. 6.68 for C-85-85. In these figures, the transient temperature distribution, shape of welded pool, out-of-plane welding distortion, and transverse plastic strain are all depicted. During the thermal elastic-plastic FE analysis for multi-pass welded joint, parallel computation with OpenMP technology was employed to enhance the computational efficiency and reduce the computational time significantly. The computing times required by thermal elastic-plastic FE analysis for each typical welded joint are reported in Table 6.13. When parallel computation technology with OpenMP was employed, the computing time for AXS-85-85 (6.55 h) was reduced to be 30% of the computing time for COVN-14-14 (23.08 h). Note also that AXS-85-85 presented many more nodes and elements than COVN-14-14; even AXS-85-85 exhibited three times the number of welding passes of COVN-14-14. In particular, the computation time for C-85-85 with 50 welding passes was 17.22 h with OpenMP, which is acceptable for engineering applications.

6.3.2.3   Inherent Deformation Evaluation and Its Validation

As mentioned above, we employed two methods to evaluate the welding inherent deformation of the examined welded joints: inverse analysis with welding distortion [19] and integration with residual plastic strain based on the definition of welding

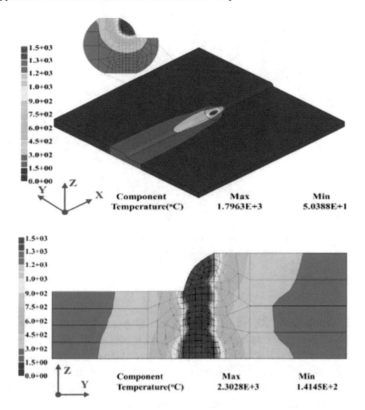

(a) Transient temperature distribution and welded pool

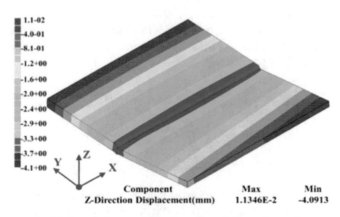

(b) Out-of-plane welding distortion (deformed rate: 5)

Fig. 6.64 Computed thermal-mechanical results of the welded joint for AI-14-22

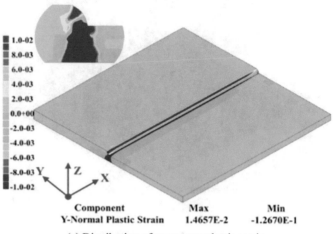

Component	Max	Min
Y-Normal Plastic Strain	1.4657E-2	-1.2670E-1

(c) Distribution of transverse plastic strain

Fig. 6.64 (continued)

inherent deformation [1, 20]. Both methods were also employed to evaluate welding inherent deformations of butt and fillet welded joints, and elastic FE analysis incorporating the evaluated welding inherent deformation was carried out to predict welding distortion. Results of elastic FE analysis with evaluated welding inherent deformation were then compared with the results of thermal elastic-plastic FE computation to confirm the precision of the evaluated inherent deformation with different methods.

In particular, longitudinal inherent shrinkage is usually converted to longitudinal shrinkage force (tendon force) according to Eq. (6.9) below during elastic FE analysis owing to strong self-constraint and non-uniform distribution of shrinkage in longitudinal direction [18, 19]. Tables 6.14 and 6.15 summarize the evaluated welding inherent deformations with both inverse analysis and the integration method for butt and fillet welded joints, respectively.

$$F_{longitudinal} = \int \varepsilon_{longitudinal}^{plastic} \times E \, dA = E \iint \varepsilon_{longitudinal}^{plastic} dy dz$$

$$= E \times h \times \frac{1}{h} \iint \varepsilon_{longitudinal}^{plastic} dy dz$$

$$= E \times h \times \delta_{longitudinal}^{inherent} \tag{6.9}$$

To validate the accuracy of the evaluated inherent deformations with inverse analysis and the integration method, butt welded joint AI-10-14 was modeled with solid elements and shell elements as shown in Figs. 6.69a and 6.70a, respectively.

According to the plotting contour of the out-of-plane welding distortions shown in Figs. 6.69b, 6.70b and c, the deformed tendency is almost identical for the three cases. Considering the points on line 1 indicated in Fig. 6.70a, the magnitudes of out-of-plane welding distortions were compared as shown in Fig. 6.71. Note that the

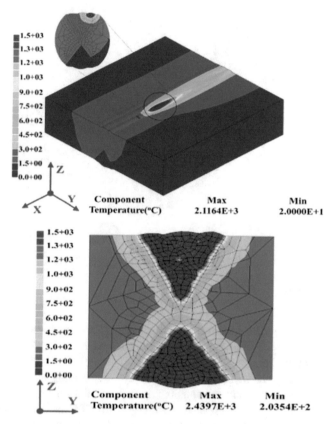

(a) Transient temperature distribution and welded pool

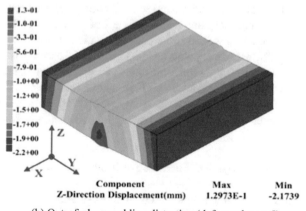

(b) Out-of-plane welding distortion (deformed rate: 5)

Fig. 6.65 Computed thermal-mechanical results of the welded joint for AXS-85-85

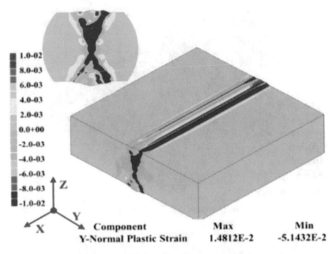

(c) Distribution of transverse plastic strain

Fig. 6.65 (continued)

out-of-plane welding distortion resulting from elastic FE computation with welding inherent deformations evaluated by inverse analysis was in good agreement with that of TEP FE computation.

A fillet welded joint was also examined as typical welding in shipbuilding for validation of welding inherent deformation. As shown in Figs. 6.72a and 6.73a, solid-element and shell-element models were employed for TEP and elastic FE computations. Computed out-of-plane welding distortions presented similar deformed tendency, as shown in Figs. 6.72b, 6.73b, and c.

For fillet welding, both flange and web presented bending deformation. Therefore, points on line 1 of the flange edge and points on line 2 of the web top edge, as shown in Fig. 6.73, were selected for welding inherent deformation validation. As shown in Fig. 6.74, bending deformations of points on line 1 in Fig. 6.73 had consistent features with each other. However, a slight difference between the bending deformations of points on line 2 computed by TEP and elastic FE computations can also be observed.

According to the above comparisons, inherent deformations evaluated by inverse analysis may be a better choice for typical welding joints in shipbuilding.

6.3.2.4   Database and Empirical Formulae of Inherent Deformation

In general, longitudinal bending is ignored owing to its smaller magnitude. Welding inherent deformations are dominantly determined by the welding heat input when the welded joint and material properties are both fixed. To preclude the influence of plate thickness, inherent deformations were uniformly considered, regardless of the plate thickness, and then converted to tendon force and parameters for all the examined

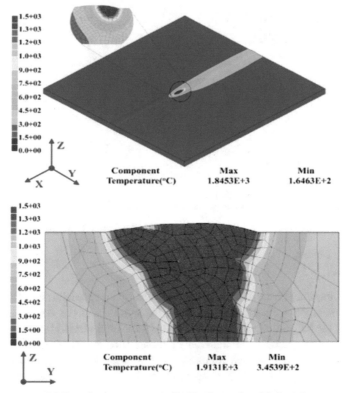

(a) Transient temperature distribution and welded pool

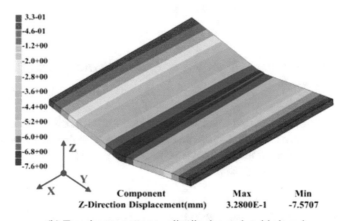

(b) Transient temperature distribution and welded pool

Fig. 6.66 Computed thermal-mechanical results of the welded joint for COVN-14-14

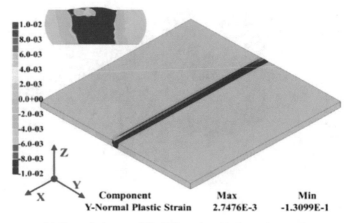

<table>
<tr><td>Component</td><td>Max</td><td>Min</td></tr>
<tr><td>Y-Normal Plastic Strain</td><td>2.7476E-3</td><td>-1.3099E-1</td></tr>
</table>

(c) Out-of-plane welding distortion (deformed rate: 5)

Fig. 6.66   (continued)

welded joints, which were calculated by multiplication with thickness. The evaluated welding inherent deformations for sequential elastic FE analysis are summarized in Table 6.16 for butt welded joints and Table 6.17 for fillet welded joints.

Based on the database summarized in Tables 6.16 and 6.17, welding inherent deformations of other welded joints in the examined ship structure can also be obtained without heavy TEP FE computation to decrease the complexity of FE analysis. Moreover, a linear regression approach was employed, as shown in Fig. 6.75, to establish the relationships between inherent deformation parameters and welding heat input. These formulae are listed in Table 6.18.

6.3.3   Prediction and Validation of Out-of-Plane Welding Distortion

With the welding inherent deformation used as load for the mechanical response of welded joints, elastic FE computation can be carried out to predict welding distortion efficiently without precision loss. This is considered a practical engineering tool for precision fabrication of large ship structures.

6.3.3.1   Watertight Transverse Bulkhead Structure

As shown in Fig. 6.76, a geometrical model of the examined watertight transverse bulkhead structure was introduced in which the plate stiffeners, L-section steel, and even the opening hole for the pipeline system continuing in the plate stiffeners were

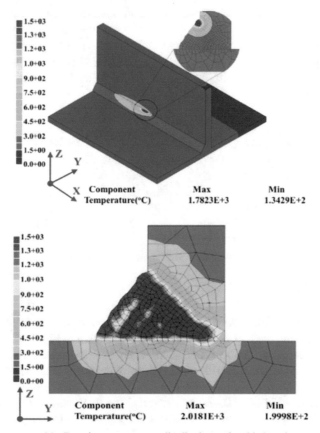

(a)   Transient temperature distribution and welded pool

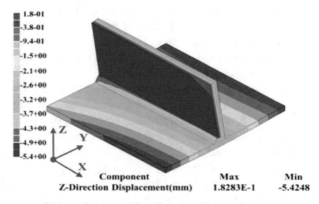

(b) Out-of-plane welding distortion (deformed rate: 5)

Fig. 6.67   Computed thermal-mechanical results of the welded joint for FP-22-16

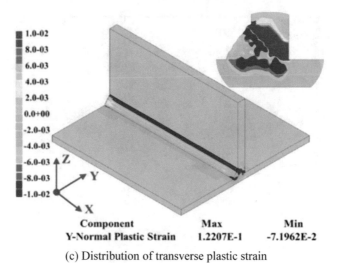

Component   Max   Min
Y-Normal Plastic Strain   1.2207E-1   -7.1962E-2

(c) Distribution of transverse plastic strain

Fig. 6.67 (continued)

all considered. In addition, the dimensional size of the skin plate, the L-section steel, plate stiffeners, and their spacing distances are also shown in Fig. 6.76.

With shell elements, identical dimensions to those mentioned above were utilized. The plate thicknesses were also identical to those of the actual ship structure. The FE mesh model was then created, as shown in Fig. 6.77. The total number of nodes and elements were 10,631 and 10,478, respectively. The total number of parts was 92; they are distinguished with different colors in Fig. 6.81. The welding lines between the adjacent parts were marked with different colors. They were used to load welding inherent deformations for elastic FE computation.

Considering the rigidity of section steels and plate stiffeners, the boundary condition shown in Fig. 6.77 was employed to prevent rigid body motion. Points on lines 1–4 were marked with different colors in Fig. 6.77 to validate the magnitude of computed welding distortions and compare them with measurements later.

Compared with in-plane shrinkage distortion, out-of-plane welding distortion exerted a much more significant influence on the dimensional precision of the ship structure. As shown in Fig. 6.78, the distribution of computed out-of-plane welding distortion was obtained. Note that deformed shape with overall middle hogging was observed and a maximal magnitude of approximately 20.87 mm appeared at the center region. This type of welding distortion may have resulted from the different plate thicknesses, density of welding line, and corresponding welding heat inputs.

As mentioned above, points on lines 1 and 2 in Fig. 6.78 were selected to validate the accuracy of the computed out-of-plane welding distortion. Figure 6.79 shows a comparison of points on lines 1 and 2 between computed out-of-plane welding distortions and experimental results. The tendency and magnitude of welding distortion

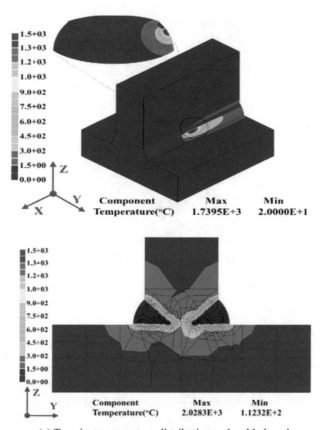

(a) Transient temperature distribution and welded pool

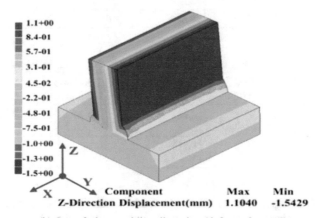

(b) Out-of-plane welding distortion (deformed rate: 5)

Fig. 6.68 Computed thermal-mechanical results of the welded joint for C-85-85

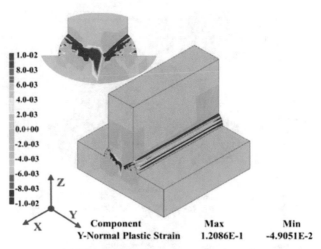

(c) Distribution of transverse plastic strain

Fig. 6.68 (continued)

Table 6.13 Comparison of computing times with OpenMP

Welded joint type	No. of nodes	No. of elements	No. of pass	OpenMP	Computing time/hour
FYS-16-16	14,433	12,150	6	No	52.23
AI-14-22	14,076	11,850	4	No	20.72
COVN-14-14	14,433	12,300	6	No	23.08
FP-22-16	21,930	18,900	16	No	161.87
AXS-85-85	16,554	15,210	18	Yes	6.55
C-85-85	32,054	29,880	50	Yes	17.22

Table 6.14 Evaluated welding inherent deformations of the butt welded joint (AI-10-14)

	Longitudinal shrinkage (mm)	Transverse shrinkage (mm)	Transverse bending (rad)	Longitudinal bending (rad)
Inverse analysis	0.075	0.265	0.0333	0.0025
Integration method		0.516	0.121	

presented good agreement with each other, while their distributions can be explained in terms of structural features and corresponding welding heat inputs.

Table 6.15 Evaluated welding inherent deformations of the fillet welded joint (FP-9-14)

	Tendon force (N)	Transverse shrinkage (mm)	Transverse bending (rad)
Inverse analysis	2.006E + 11	0.199 (flange) 0.229 (web)	0.0085 (flange) 0.0416 (web)
Integration method		0.126 (flange) 0.286 (web)	

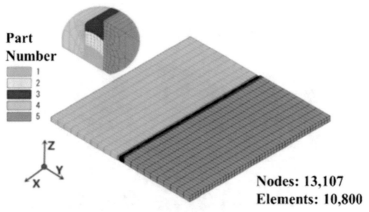

Nodes: 13,107
Elements: 10,800

(a) Solid-element model of the butt welded joint

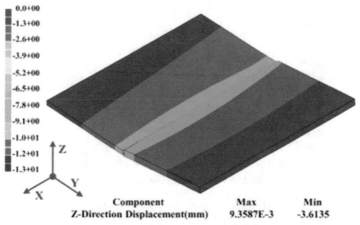

Component   Max   Min
Z-Direction Displacement(mm)   9.3587E-3   -3.6135

(b) Computed out-of-plane welding distortion

Fig. 6.69 Solid-element model and computed out-of-plane welding distortion of the butt welded joint (AI-10-14)

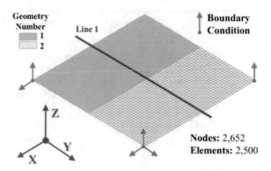

(a)   Shell-element model of the butt welded joint

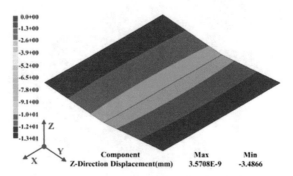

(b) Out-of-plane welding distortion with inherent deformations evaluated by inverse analysis

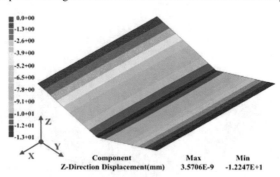

(c) Out-of-plane welding distortion with inherent deformations evaluated by the integration method

Fig. 6.70 Shell-element model and out-of-plane welding distortion of the butt welded joint with elastic FE analysis (AI-10-14)

Fig. 6.71 Comparison of butt welding distortion of TEP FE analysis and elastic FE computation with inherent deformations by different evaluation approaches

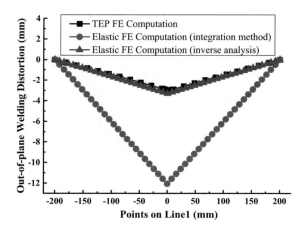

6.3.3.2   Torsion Box Structure

Given that FE mesh influences the computed accuracy and computer consumption, ship structure modeling as engineering application should ensure identical geometrical profiles and utilize a fine mesh for the region near the welding line.

According to the actual torsion box structure, Fig. 6.80 shows the geometrical dimensions; 60 parts are marked with different colors. The FE mesh with shell elements is depicted in Fig. 6.81, in which the number of nodes and elements are 10,554 and 8,741, respectively. The welding lines were automatically generated between the adjacent parts with different colors. Boundary condition considering structural stiffness was employed to prevent rigid body motion, as shown in Fig. 6.81. Lines 1–4 were also marked and selected for computed accuracy validation by comparing predicted out-of-plane welding distortion with the corresponding measurements.

Figure 6.82 shows the plotting contour of the computed out-of-plane welding distortion with a deformed rate of 5. Note that minimal out-of-plane welding distortion with 7.00-mm magnitude appeared at the middle region due to the angular distortion resulting from welding of thick plates welding with 85 mm and 60 mm in thickness. Owing to the weak stiffness, a maximal out-of-plane welding distortion with magnitude of 9.68 mm was generated at the edge region.

To compare the computed results accurately, the welding distortion of points on lines 1 and 2 in Fig. 6.81 was selected and compared with measurements, as shown in Fig. 6.83. Good agreement can be clearly observed not only for deformed tendency but also for magnitudes.

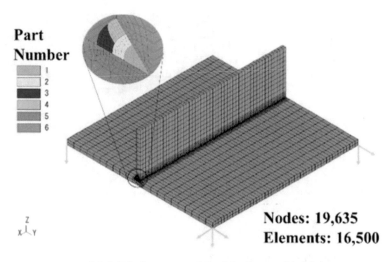

(a) Solid-element model of the butt welded joint

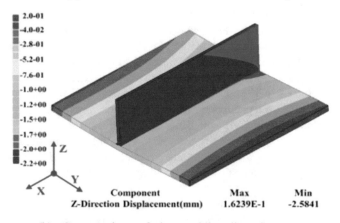

(b)   Computed out-of-plane welding distortion

Fig. 6.72   Solid-element model and computed out-of-plane welding distortion of the butt welded joint (FP-9-14)

6.3.4   Influence of Welding Sequence on Precision Fabrication

For complicated welded structures with many parts, the final dimensional precision was determined not only by the mechanical response of welded joints but also by the welding sequence and fitting configuration.

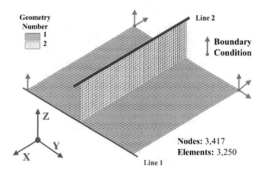

(a) Shell-element model of the fillet welded joint

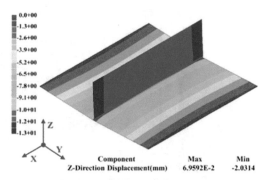

(b) Out-of-plane welding distortion with inherent deformations evaluated by inverse analysis

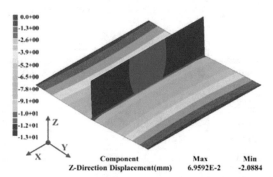

(c) Out-of-plane welding distortion with inherent deformations evaluated by the integration method

Fig. 6.73 Shell-element model and out-of-plane welding distortion of the fillet welded joint with elastic FE analysis (FP-9-14)

Fig. 6.74 Comparison of fillet welding distortion of TEP FE analysis and elastic FE computation with inherent deformation by different evaluation approaches

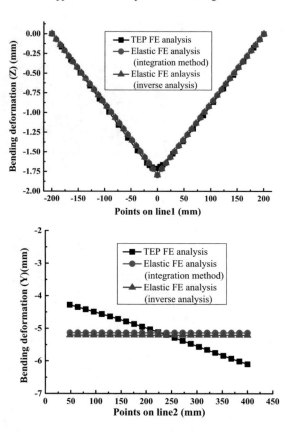

Table 6.16 Evaluated inherent deformations for examined butt welded joints

Weld joints code	Heat input (J/mm)	Tendon force (N)	Transverse Shrinkage*Thickness (mm^2)	Transverse Bending*Thickness3 (rad*mm^3)
FYS-12-12	2868.2	8.90E+10	4.82	83.12
FYS-12-16	3539.3	1.61E+11	6.81	178.36
FYS-16-16	4478.8	2.09E+11	8.16	287.13
COVN-10-10	2215.0	1.71E+11	1.50	52.30
AI-10-14	2347.3	1.85E+11	3.18	57.54
AI-14-22	3146.6	2.22E+11	4.99	222.78
AXS-85-85	27236.1	2.04E+12	69.67	5957.01
AXS-60-60	15289.5	9.57E+11	29.04	1922.40
AXS-85-60	15289.5	1.06E+12	37.98	2057.82
COXS-85-85	73210.6	1.59E+12	169.70	29907.89
COXS-40-40	25025.0	5.83E+11	15.17	768.00
COVN-14-14	3755.0	1.38E+11	4.73	99.06
AYN-15-18	1961.2	1.58E+11	4.49	74.57

Table 6.17   Evaluated inherent deformations for examined fillet welded joints

Weld joints code	Heat input(J/mm)	Tendon force(N)	P1	P2	P3	P4
FP-9-14	2379.9	2.01E +11	2.8	2.1	23.3	30.3
FP-10-13	2629.9	2.08E +11	2.6	1.3	20.4	38.5
FP-13-16	4388.1	2.74E +11	4.3	1.2	32.8	128.3
FP-22-16	8712.4	4.68E +11	7.4	7.3	69.6	890.2
FP-15-60	5571.5	2.07E +11	7.8	23.5	799.2	249.4
FP-20-70	8078.6	2.74E +11	10.1	41.1	1200.5	672.8
PP-80-80-15	9655.9	6.79E +11	11.6	26.4	1587.2	2764.8
PP-80-80-20	17494.0	1.05E +12	12.3	35.5	921.6	4147.2
PP-80-80-27	27451.5	1.26E +12	14.3	44.8	1126.4	8908.8
PP-85-85-15	9655.9	6.72E +11	13.0	0.4	1412.5	2333.7
PP-85-85-18	14350.5	7.10E +11	12.2	13.1	736.95	4298.9
PP-85-85-20	17494.0	9.97E +11	16.5	6.6	1105.4	3869.0
PP-85-85-24	21475.8	8.29E +11	14.8	55.4	61.4	9826.0
PP-85-85-33	44226.0	1.53E +12	15.3	97.4	1842.4	17748.

P1: Flange Transverse Shrinkage *Thickness (mm^2); P2: Web Transverse Shrinkage*Thickness (mm^2); P3: Flange Transverse Bending*Thickness3 (rad*mm^3); P4: Web Transverse Bending*Thickness3 (rad*mm^3)

The influence of the welding sequence was examined in the fabrication of the watertight transverse bulkhead structure by elastic FE computation. For actual fabrication in shipbuilding, ship blocks such as the examined watertight transverse bulkhead structure of an ultra-large container would be sequentially assembled from steel pieces, parts, and segments. Welding distortion during steel pieces joining can be significantly reduced with the application of many mitigation techniques. Thus, the dimensional precision of segments can be guaranteed. However, blocking the welding distortion during the fabrication of segments may be difficult to control owing to the large structure stiffness of assembled segments.

The watertight transverse bulkhead structure was selected to examine the influence of the welding sequence segments on dimensional precision. As shown in Fig. 6.84, the watertight transverse bulkhead structure was assembled from two segments (TB11A and TB11B) with conventional fabrication. In addition, TB11A and TB11B were fabricated separately, and then were assembled together for the sake of high productivity. However, the TB11B segment is too large to control and reduce the welding distortion during its fabrication. It is affected by the fitting precision and final dimensional quality when it is assembled with the B11A segment. Therefore, it is better to separate the examined watertight transverse bulkhead structure to be five segments for fabrication, as shown in Fig. 6.85.

Considering structural continuity in the context of the proposed fabrication plan, the welding distortion can be easily reduced during the process of joining each segment and during the process of assembling them with some extra work hours.

Fig. 6.75 Formulae of
welding inherent
deformations from linear
regression analysis

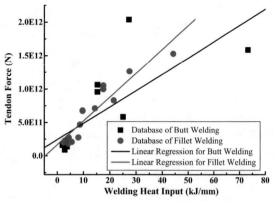

(a) Linear regression for tendon force formula

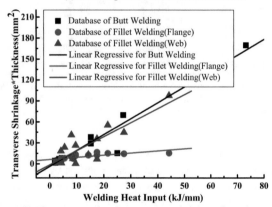

(b) Linear regression for transverse shrinkage formula

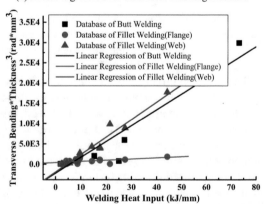

(c) Linear regression for transverse bending formula

Table 6.18 Evaluated formulae relating inherent deformations to welding heat input

Parameter meaning	Butt welded joint	Fillet welded joint
Tendon force	$2.43 \times 1011 + 2.44 \div 1010 \times Q_{net}$	$1.66 \times 1011 + 3.56 \div 1010 \times Q_{net}$
Transverse Shrinkage*Thickness	$-3.85 + 2.27 \times Q_{net}$	$flange : 6.11 + 0.31 \times Q_{net}$ $web : Y = -1.54 + 2.01 \times Q_{net}$
Transverse Bending*Thickness	$-2194.86 + 389.22 \times Q_{net}$	$flange : 399.6 + 29.14 \times Q_{net}$ $web : -1925.69 + 429.49 \times Q_{net}$

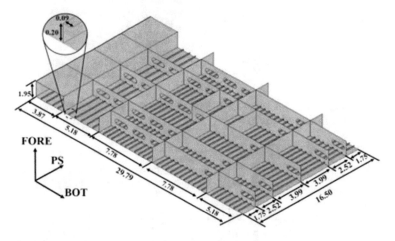

Fig. 6.76 Geometrical model of the examined watertight transverse bulkhead structure (BOT: ship-depth direction; PS: ship-side direction; FORE: ship bow and stern direction) (unit: mm)

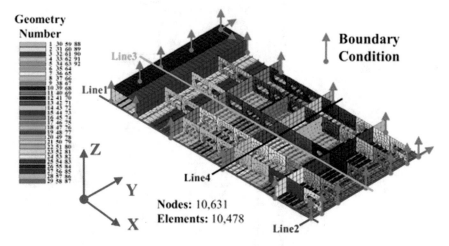

Fig. 6.77 Shell-element model of the watertight transverse bulkhead structure

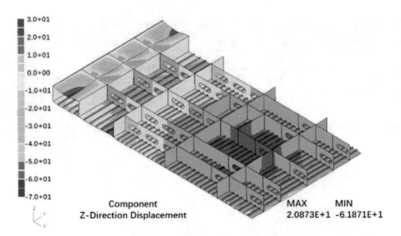

Fig. 6.78 Plotting contour of out-plane welding distortion of examined watertight transverse bulkhead structure (deformed rate: 5)

Fig. 6.79 Comparisons of out-of-plane welding distortions of examined points from measurements and FE computation (watertight transverse bulkhead structure)

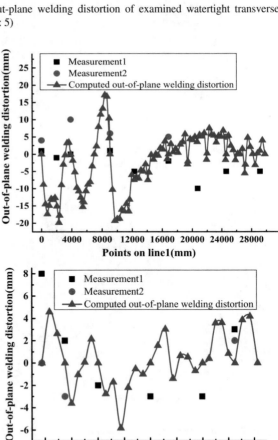

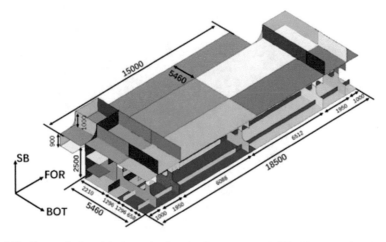

Fig. 6.80 Geometrical model of examined torsion box structure (BOT: ship-depth direction; SB: ship-starboard direction; FORE: ship bow and stern direction) (unit: mm)

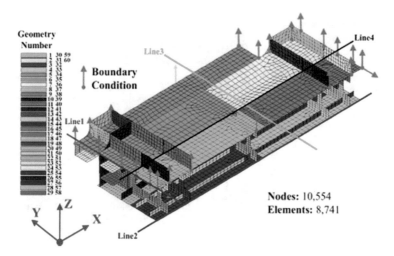

Fig. 6.81 Shell-element model of the torsion box structure

Moreover, the requirement of lifting capacity can be reduced, and the coefficient of workshop utilization and even the shipway can be clearly improved. The weight of the segment of the TB11B structure in the conventional fabrication plan was approximately 71,190 kg. Among all the segments in the proposed fabrication plan, the largest segment was the 3rd structure. Its weight was reduced by 43.7%, becoming 40,081 kg, which is convenient for utilization of lifting equipment. The segments, their parts, and weights for both the conventional and proposed fabrication plans are summarized in Table 6.19.

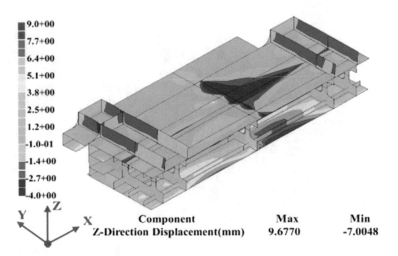

Fig. 6.82 Plotting contour of out-of-plane welding distortion of the examined torsion box structure (deformed rate: 5)

Fig. 6.83 Comparison of out-of-plane welding distortions of examined points from measurement and FE computation (torsion box structure)

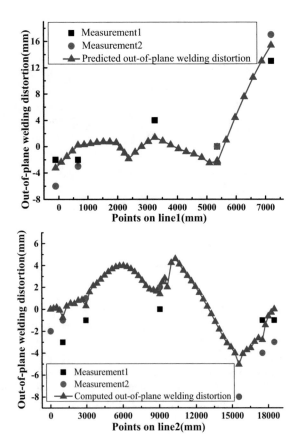

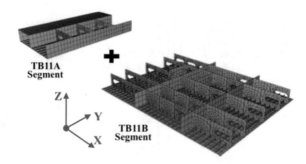

Fig. 6.84 Assemble plan during conventional fabrication of the examined watertight transverse bulkhead structure

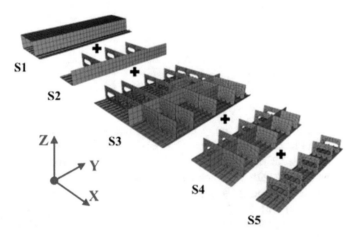

Fig. 6.85 Proposed assemble plan during the fabrication of examined watertight transverse bulkhead structure

Table 6.19 Comparison of conventional and proposed fabrication plans

Conventional			Proposed				
Segment	TB11A	TB11B	S1	S2	S3	S4	S5
No of Parts	26	66	17	9	36	20	10
Weight (kg)	36425	71190	25177	11248	40081	20654	10454

According to the proposed segments for the examined watertight transverse bulkhead structure listed in Table 6.19, the overall influence of their welding sequence on out-of-plane welding distortion was then examined. As summarized in Table 6.20, four different welding sequences of sequential assembly were considered, namely conventional welding, sequential welding, symmetrical welding, and intermittent welding. In addition, conventional welding included two segments, so-called TB11A and TB11B, as shown in Fig. 6.84, which were directly welded together. Moreover,

Table 6.20 Comparison of welding distortion with different welding sequences

Welding sequence	No. of segment	Overall maximal out-of-plane welding distortion (mm)	Relative deflection of points on line 1 (mm)	Relative deflection of points on line 2 (mm)	Total relative deflection (mm)
Conventional welding	2	20.87	1478.00	365.00	1842.98
Sequential welding	5	17.90	916.61	292.62	1209.24
Symmetrical welding from center to edge	5	17.43	791.95	304.54	1096.49
Symmetrical welding from edge to center	5	17.97	963.40	292.47	1255.87
Intermittent welding	5	19.53	1177.47	406.01	1583.48

there were five components in the examined watertight transverse bulkhead structure, as shown in Fig. 6.85, which were called S1-S5. Sequential welding means that S1 was welded with S2, S2 was then welded with S3, S3 was welded with S4, and finally S4 was welded with S5. Symmetrical welding means that S1 and S2 were welded simultaneously and S4 and S5 were also welded simultaneously, then S4 and S2 were welded with S3 simultaneously. Intermittent welding means that S1 was welded with S2 and S3 was then welded with S4; subsequently, S2 was welded with S3, and S4 was welded with S5. The overall maximal out-of-plane welding distortions are summarized in Table 6.20.

According to these different welding sequences, out-of-plane welding distortions of points on lines 3 and 4 shown in Fig. 6.81 are depicted in Fig. 6.86. Note that the out-of-plane welding distortion can be significantly reduced with the proposed fabrication plans. However, their distributions cannot be easily compared to analyze which welding sequence was the optimal case quantitatively. Therefore, relative deflection is proposed, which is defined as follows:

$$relative\ deflection = \sum_{i=1}^{Point\ No.} |deflection(i)| \qquad (6.10)$$

Relative deflections of points on lines 1 and 2 shown in Fig. 6.81 are summarized in Table 6.20. Among the considered welding sequences, it can be concluded that the best welding sequence to reduce the out-of-plane welding distortion of the examined watertight transverse bulkhead structure was symmetrical welding from center to edge. This can be explained in terms of the stiffness prior theory (SPT). The total

Fig. 6.86 Comparison of out-of-plane welding distortion with different welding procedures

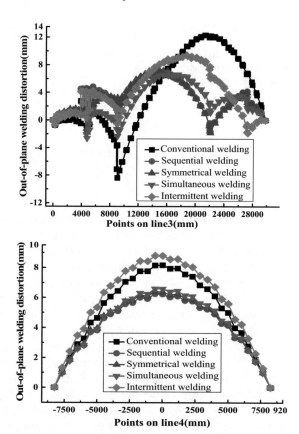

relative deflection by symmetrical welding from center to edge was approximately 1096.49 mm, which means a reduction of approximately 40% from 1842.98 mm achieved by conventional welding. Its maximal out-of-plane welding distortion was approximately 17.43 mm, which means a reduction of approximately 20% from 20.87 mm.

6.3.5   X Groove Optimization of Butt Welded Joint with Thick Plates

Currently in shipbuilding, two practical techniques are employed for out-of-plane welding distortion mitigation during butt welding of thick plates with X type groove: symmetrical welding and inverse deformation application [20]. However, both techniques have engineering limitations. In particular, symmetrical welding requires frequent turnover of welded joint during the assemble procedure, which significantly reduces the fabrication efficiency, requires much more equipment, and even

Fig. 6.87 Groove profile of butt welding with thick plates (48 < t ≤ 100) (h1: depth of upper welding; h2: depth of lower welding)

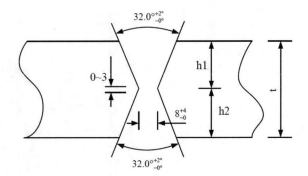

increases the probability of accident. Inverse deformation application can ensure the welding productivity without many turnovers of welded joint, but the magnitudes of inverse deformations are not easy to obtain because it requires a great deal of experimental measurements or TEP FE computations owing to their dependence on material properties and plate thickness. Moreover, the mold weight may require to support mechanical constrains during the inverse deformation application.

Therefore, the design optimization of welding groove with X type was examined. In addition, out-of-plane welding distortion generated by upper welding can be reduced by lowering the welding in the butt welding of thick plates with asymmetrical groove. This is advantageous for high productivity, giving rise to less out-of-plane welding distortion compared with current mitigation practices.

Actually, symmetrical groove design for butt welding with thick plates, shown in Fig. 6.87, was employed. The groove angle was approximately 32°, the gap was approximately 8 mm, and the length of the truncated edge was approximately 3 mm. According to experimental measurements, out-of-plane welding distortion using the above mentioned groove design extended the fabrication tolerance. Concerning the upper welding passes, they were finished and then turnover the welding joint took place once for lower welding passes.

The underlying mechanism is almost the same as for transverse plastic strains generated during both upper and lower welding. However, the bending moment caused by transverse plastic strains and structural stiffness was altered during the entire welding procedure with wire fusion and filling.

Asymmetrical X groove can provide a particular depth ratio of upper and lower welding passes defined by Eq. (6.11) below to mitigate out-of-plane welding distortion with only one turnover. Moreover, it should be ensured that the groove angle is not smaller than 32° as a result of actual welding procedures:

$$depth\,ratio = \frac{h1(depth\ of\ upper\ welding)}{h2(depth\ of\ lower\ welding)} \qquad (6.11)$$

Taking the butt welding of a thick plate with 60 mm in thickness as an example, Fig. 6.88 shows the solid-element model and a series of TEP FE computations that were carried out to optimize the asymmetrical groove design with less out-of-plane

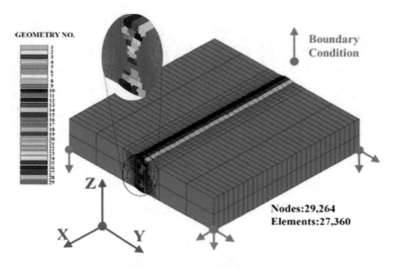

Fig. 6.88 Solid-element model of the butt welded joint (thickness: 60 mm)

welding distortion. The considered solid-element model was 300 mm in length, 300 mm in width, and 60 mm in thickness. The total number of elements and points were 27,360 and 29,264, respectively.

To improve the computational efficiency, an in-house code for the auto bead technique was proposed [21] and employed to confirm the number and position of welding pass for asymmetrical groove design with different depth ratios. The cross-section areas of each welding pass were identical owing to the application of the same welding conditions. One solid-element model was required in which a fine mesh was created near the welding region and a coarse mesh was used for the region far away of the welding pass.

Geometrical features of welding passes can be described by origin coordinates, width, depth, and height of the heated region with a double ellipsoid model. Then, groove design can be subsequently modified according to the arrangement of welding passes shown in Fig. 6.89.

We first examined the case in which the depth of the upper welding groove ($h1$) equaled the depth of the lower welding groove ($h2$). The case in which the depth ratio is one was also examined. According to the SAW welding condition in Table 6.12 and its corresponding deposition amount, an arrangement of welding passes for butt welding with 60-mm thick plates was designed, as shown in Fig. 6.89. Essentially, welding passes in the lower groove started when all the welding passes in the upper groove finished and turnover of the welded joint had already occurred.

Thermal analysis was carried out with in-house TEP FE programming. Then, its computed temperature profile was applied as a thermal load to examine the mechanical response. The distribution of maximal temperature and the geometrical profile of the molten pool are shown in Fig. 6.90. In addition, the plotting contour of the final out-of-plane welding distortion in cross-section view is depicted in Fig. 6.90.

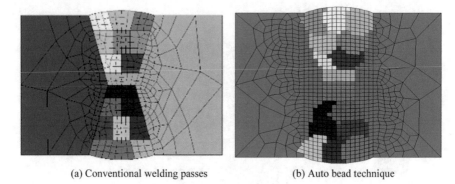

(a) Conventional welding passes   (b) Auto bead technique

Fig. 6.89 Comparison of the arrangement of welding passes between conventional and proposed methods

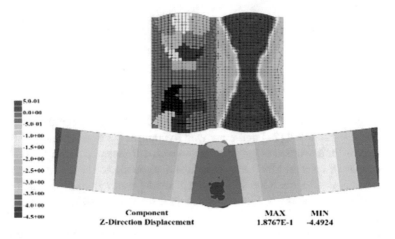

Fig. 6.90 TEP FE computed results of the butt welding (depth ratio: 1/1)

Note that the predicted geometrical profile of welding molten pool agrees with the dimension of the designed groove. Note also the out-of-plane welding distortion with approximately 4.5-mm magnitude. This distortion is not easy to mitigate. Therefore, it can be concluded that the welding groove when the depth ratio was set to one was not appropriate to reduce the out-of-plane welding distortion.

As mentioned above, the bending moment caused by welding plastic strain in transverse direction and the structural stiffness due to filler deposition were both developed all the welding procedures. They jointly determined the eventual out-of-plane welding distortion. Moreover, a series of TEP FE computations with different depth ratios were carried out. The distribution of maximal temperature and plotting contour of out-of-plane welding distortion is shown in Fig. 6.91. The squeeze theorem for domain approximation was employed and implemented as shown in Fig. 6.92.

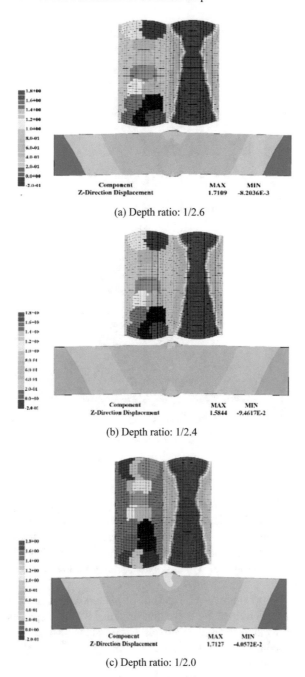

(a) Depth ratio: 1/2.6

(b) Depth ratio: 1/2.4

(c) Depth ratio: 1/2.0

Fig. 6.91 TEP FE computed results of the butt welding with different depth ratios

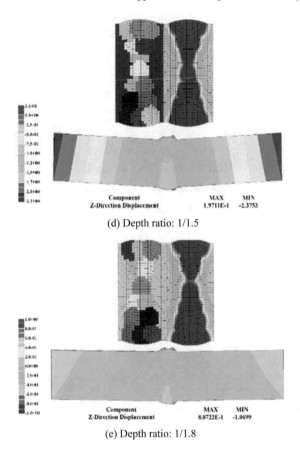

(d) Depth ratio: 1/1.5

(e) Depth ratio: 1/1.8

Fig. 6.91 (continued)

In addition, if the welded joint bends downward with negative magnitude of out-of-plane welding distortion, the depth of lower groove should be increased; otherwise, the depth of lower groove should be decreased with upward bending deformation. Table 6.21 summarizes the computed results with different depth ratios, such as temperature profile, number of welding passes, maximal deflection, and flatness of top surface.

The computed temperature profile and out-of-plane welding distortions are shown in Fig. 6.93 for a depth ratio of 1/1.9. In addition, 5 and 8 welding passes were applied for the upper and lower grooves, respectively. Note that the final out-of-plane welding distortions were reduced below 1.0 mm due to the effect of transverse plastic strain generated by lower welding.

Both the magnitude of out-of-plane welding distortion and flatness of top surface were employed to evaluate the welding groove, as summarized in Table 6.21. The flatness of the top surface can be explained according to the value of the welding

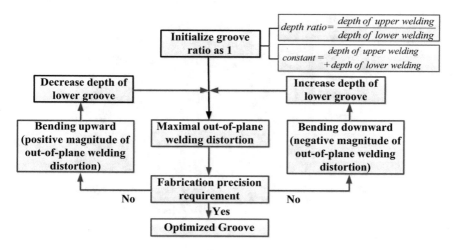

Fig. 6.92 Programming chart for optimized groove search

Table 6.21 Comparison of computed results with different depth ratios

Depth ratio	Maximal temperature (°C)	Minimal temperature (°C)	Welding pass (upper/lower)	Maximal out-of-plane welding distortion (mm)	Flatness of top surface (°)
1:1.0	2113	207	8/8	4.50	3.44
1:2.6	2208	204	3/10	−1.71	−1.32
1:2.4	2275	204	3/10	−1.58	−1.2
1:2.0	2200	199	5/10	−1.71	−1.32
1:1.5	1994	206	6/8	2.38	1.82
1:1.8	2040	201	5/8	1.07	0.82
1:1.9	1987	206	5/8	0.88	0.68

angle with respect to the initial configuration and the welding inherent transverse bending.

Moreover, the points on cross-section line of the bottom surface were examined. Figure 6.94 shows a comparison of the out-of-plane welding distortions for different depth ratios. It can be clearly seen that a smaller magnitude of out-of-plane welding distortion was generated when the depth ratio was 1/1.9 compared with the other depth ratios. A large magnitude of out-of-plane welding distortion was generated by upper welding with the typical groove design. It could not be completely reduced by the lower welding and its generated bending moment resulting from strong structural stiffness.

The point of truncated edge was also examined. For comparison of the welding procedure of the typical groove design with that of optimized groove, Fig. 6.95 shows

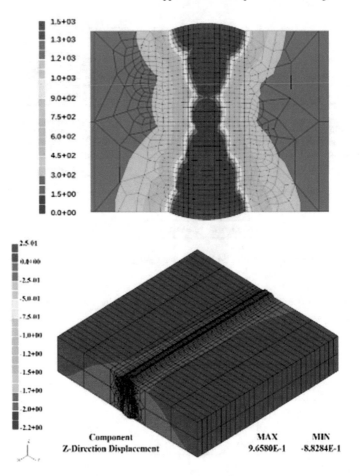

Fig. 6.93 TEP FE computed results of butt welding with optimized groove (depth ratio: 1/1.9,deformed rate: 5)

the evolution of out-of-plane welding distortion of the examined point with different groove designs. Out-of-plane welding distortion could be well controlled with the optimized groove design, as shown in Fig. 6.95. Then, the TEP FE computational analysis above described was carried out to examine groove optimization for butt welding with different plate thicknesses. Table 6.22 summarizes the optimized depth ratios for plate thickness values of 40, 50, 60, 65, 70, and 85 mm. All of them utilize the same welded joint type shown in Fig. 6.87.

Based on the results summarized in Table 6.22, the relation between optimized depth ratio and plate thickness can be evaluated through regression analysis. As shown in Fig. 6.96, non-linear numerical fitting with Boltzmann function can be employed. Welding joints with smaller and larger thicknesses approximately exhibited constant depth ratio of optimized groove resulting from the balance of welding

Fig. 6.94 Comparison of out-of-plane welding distortion with different depth ratios (thickness: 60 mm)

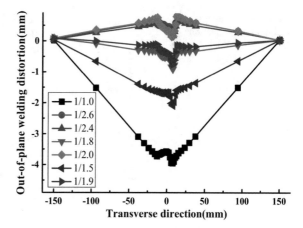

Fig. 6.95 Evolution of out-of-plane welding distortion of truncated edge with welding procedure

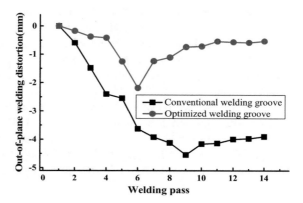

Table 6.22 Depth ratio of optimized groove for different plate thicknesses

Thickness (mm)	Optimized depth ratio
40	1:1.80 (0.556)
50	1:1.85 (0.541)
60	1:1.90 (0.526)
65	1:2.20 (0.455)
70	1:3.50 (0.286)
85	1:4.00 (0.250)

bending moment and structural stiffness. With the empirical formula, the depth ratio of optimized groove can be conveniently obtained for other thicknesses to mitigate the out-of-plane welding distortion. This improved the investigation efficiency without consuming too many computer resources.

From the database of welding inherent deformation of butt welding summarized in Tables 6.16 and 6.18, the welding groove of thick plates (40–85 mm) could be

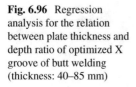

Fig. 6.96 Regression
analysis for the relation
between plate thickness and
depth ratio of optimized X
groove of butt welding
(thickness: 40–85 mm)

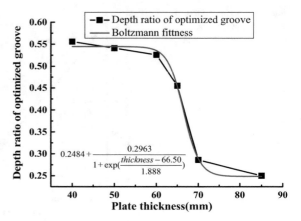

optimized with less out-of-plane welding distortion. The corresponding transverse
bending was also reduced. Applying the latest welding inherent deformations as
loading to be the shell-element model of the examined torsion box structure, elastic
FE analysis was carried out to predict fabrication precision. The resulting plotting
contour of out-of-plane welding distortion is shown in Fig. 6.82. Compared with the
computed results shown in Fig. 6.97, the magnitude of maximal out-of-plane welding
distortion was reduced by approximately 28%, from 16.68 mm to 12.03 mm.

Moreover, points on lines 3 and 4 shown in Fig. 6.81 are examined. Their out-of-
plane welding distortion with current and optimized groove designs are compared
in Fig. 6.98. Note that the magnitudes of out-of-plane welding distortions were
significantly reduced when the optimized groove design was employed.

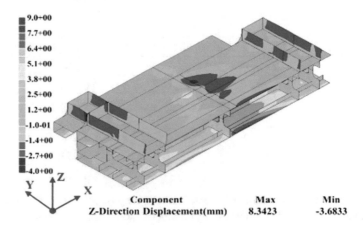

Fig. 6.97 Plotting contour of out-of-plane welding distortion considering butt welding with
optimized groove

Fig. 6.98 Comparison of out-of-plane welding distortions of examined points with different groove designs

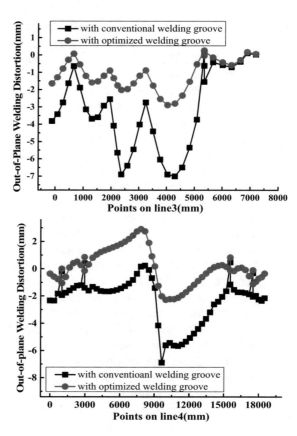

6.4 Conclusions

According to the investigation results above reported, hatch coaming production with consideration of the influence of the welding pass sequence was examined. An improved fabrication assembly was proposed and numerically considered to reduce the dimensional tolerance due to welding process. Moreover, welding induced buckling in a ship panel structure was investigated by a proposed FE computation approach in which the magnitude of inherent deformations of typical welded joints were evaluated from the previously computed results of TEP FE analysis. Elastic FE analysis based on inherent deformation theory was employed. The large deformation theory was essential to consider geometrical non-linear response. Welding distortion prediction and mitigation for watertight transverse bulkhead and torsion box structures of ultra-large container ships were demonstrated. Influences of welding sequence and groove design on out-of-plane welding distortion were numerically considered. A significant mitigation effect was observed when comparing the numerical results with the actual fabrication procedure. Overall, the following conclusions can be drawn:

(1)   An iterative substructure method and OpenMP computation were employed
       to enhance the computational efficiency during transient non-linear TEP FE
       analysis of typical welded joints. Good agreement between computed results
       and measurements can be observed.

(2)   For the assembly of the hatch coaming to the deck structure, inherent defor-
       mation was used in an elastic FE analysis to predict the top flange deflection
       resulting from welding while considering the varying gap between the hatch
       coaming and the deck. Deflection of the top flange was found to be smaller
       when using the improved welding pass sequence and well below the allowed
       tolerance required by the hatch cover.

(3)   In-plane inherent shrinkages were closely examined as the dominant reason
       of welding induced buckling. We compared their distribution and magni-
       tude in case of finishing the 1st and 2nd welding passes during sequential
       welding. Significant differences of inherent deformations between sequential
       and simultaneous welding were pointed out.

(4)   Using the small and large deformation theorems, significant differences
       between computed results in deformed mode and magnitude of out-of-plane
       welding distortion were observed. Buckling deformation was reproduced when
       large deformation theory was employed.

(5)   An integration method with inherent deformation and inverse analysis with
       welding displacement were both proposed and employed for evaluation of
       welding inherent deformation. Their accuracy was validated by elastic FE
       analysis with the shell-element model for database establishment in future
       engineering applications.

(6)   Elastic FE analysis of examined ship structures was carried out for welding
       distortion prediction with lower computer consumption. In this analysis,
       welding inherent deformation was applied as mechanical load. Good agreement
       between computed results and measurements were obtained. It was demon-
       strated that the proposed FE computation is an ideal and practical investigation
       approach to predict welding induced buckling for large welded structures, such
       as ship panels.

(7)   For the examined watertight transverse bulkhead structure, a welding sequence
       of five individual segments was numerically examined. Symmetrical welding
       from center to edge can significantly reduce the welding distortion compared
       with a conventional welding procedure. Maximal and total relative deflections
       were reduced by approximately 20 and 40% respectively owing to the structural
       stiffness reinforcement with the proposed welding sequence.

(8)   Welding groove optimization with asymmetrical feature of butt welding thick
       plate (40–85 mm) was considered for fabrication precision control of the exam-
       ined torsion box structure. This was advantageous in terms of high productivity
       and low cost. Optimized groove depth ratios of different plate thicknesses
       with highly effective TEP FE computation were calculated and corresponding
       welding inherent deformations were obtained for elastic FE analysis. Out-of-
       plane welding distortion could be significantly reduced by approximately 28%
       with the optimized groove design for butt welding thick plate.

References

1. Wang JC, Ma N, Murakawa H (2015) An efficient FE computation for predicting welding induced buckling in production of ship panel structure. Mar Struct 41:20–52
2. Eyres DJ, Bruce GT (2012) Decks, hatches and superstructures. In: Ship Construction, 7th ed., Butterworth-Heinemann, Kidlington, Oxford, UK, pp 225–240
3. Verhaeghe G (1999) Predictive formulate for weld distortion-a critical review. Abington Publishing, Cambridge, UK
4. Masubuchi K (1980) Analysis of welded structures. Pergamon Press, Oxford, UK
5. Lee DJ, Kim, GG, Shin SB (2007) Development of control technology for global bending distortion of hatch cover in container carrier during fabricating process. In: Proceedings, 17th international off-shore and polar engineering conference, July 1–6, Lisbon, Porgugal, pp 3787–3791
6. Wang JC, Shibahara M, Zhang XD, Murakawa H (2012) Investigation on twisting distortion of thin plate stiffened structure under welding. J Mater Process Technol 212(8):1705e15
7. Liang W, Sone S, Tajima M, Serizawa H, Murakawa H (2004) Measurement of inherent deformation in typical weld joint using inverse analysis (part 1): inherent deformation of bead on welding. Trans JWRI 33(1):45e51
8. Liang W, Deng D, Sone S, Murakawa H (2005) Prediction of welding distortion by elastic finite element analysis using inherent deformation estimated through inverse analysis. Weld Word 49(11e12):30e9
9. Wang JC, Rashed S, Murakawa H, Luo Y (2013) Numerical prediction and mitigation of out-of-plane welding distortion in ship panel structure by elastic FE analysis. Mar Struct 34:135e55. (Dec 2013)
10. White JD, Leggatt RH, Dwight JB (1980) Weld shrinkage prediction. Weld Metal Fabr 1980(11):587e96
11. Wang JC, Rashed S, Murakawa H (2014) Mechanism investigation of welding induced buckling using inherent deformation method. Thin-Walled Struct 80:103e19. (July 2014)
12. Wang JC, Sano M, Rashed S, Murakawa H (2013) Reduction of welding distortion for an improved assembly process for hatch coaming production. J Ship Prod Des 29(4):1e9. (Nov 2013)
13. Ueda Y, Murakawa H, Ma N (2007) Computational approach to welding deformation and residual stress. Sanpo Publication, Japan Tokyo
14. Tajiama Y, Rashed S, Okumoto Y, Katayama Y, Murakawa H (2007) Prediction of welding distortion and panel buckling of car carrier decks using database generated by FEA. Trans JWRI 36(1):65e71
15. Wang JC, Ma N, Murakawa H (2015) An efficient FE computation for predicting welding induced buckling in production of ship panel structure. Mar Struct 41:20e52. (April 2015)
16. Ma N, Wang J, Okumoto Y (2016) Out-of-plane welding distortion prediction and mitigation in stiffened welded structures. Int J Adv Manuf Technol 84(5–8):1371–1389
17. Wang J, Zhao H, Zou J, Zhou H, Wu Z, Du S (2017) Welding distortion prediction with elastic FE analysis and mitigation practice in fabrication of cantilever beam component of jack-up drilling rig. Ocean Eng 130:25–39
18. Wang JC, Zhou H, Zhao H, Zhou FM, Ma N (2017) Comparative study on evaluation of tendon force for welding distortion prediction in thin plate fabrication. China Welding (English Edition) 26(3):1–11
19. Zhou H, Wang J (2019) Accurate FE computation for out-of-plane welding distortion prediction of fillet welding with considering self-constraint. J Ship Prod Des. https://doi.org/10.5957/jspd.03180006
20. Wang JC, Shi XH, Zhou H, Liu JF (2019) Controlling on out-of-plane welding distortion of thick plates butt welded joint in container ship. J Harbin Eng Univ. https://doi.org/10.11990/jheu.201805117
21. Goldak J, Bibby M, Moore J, House R, Patel B (1986) Computer modeling of heat flow in welds. Metall Trans B 17(3):587–600

Chapter 7
Application of Accurate Fabrication of Offshore Structure

The wide offshore exploration and development, in particular search and extraction of offshore crude oil and natural gas in the past several decades, led to progressive development and use of jack-up drilling rig to become the most popular type of mobile offshore drilling unit owing to its flexible characteristics in service. Jack-up drilling rigs, which are bottom supported units, usually consist of a triangular hull, three legs, and a jacking system. They rest on the sea floor rather than float, and the core of a jack-up drilling rig is that it is self-elevating with three movable legs. These legs are positioned on the ocean floor and drilling equipment is jacked up above the water surface. Thus, two major components of offshore structures, namely cylindrical leg structures and cantilever beam structures, were examined with computational welding mechanics for welding distortion prediction and mitigation. The application of an advanced computational approach for accurate fabrication of offshore structures was also demonstrated.

7.1 Welding Distortion Prediction and Mitigation Practice of Cylindrical Leg Structure

In general, there are two typical types of leg structures during offshore structure fabrication: truss and cylinder forms. In addition, a cylindrical leg structure as the main component of jack-up rig with frequent self-elevation plays an essential role for stiffness support and safety operation of a working platform. Its fabricated dimensional accuracy significantly influences the operation cost and service life of the whole jack-up rig. Therefore, it is a critical engineering problem to control and prevent the welding deformation between the rack and the cylinder during the fabrication of cylindrical leg structures.

© Science Press 2021
H. ZHOU and J. WANG, *FE Computation on Accuracy Fabrication of Ship and Offshore Structure Based on Processing Mechanics*,
https://doi.org/10.1007/978-981-16-4087-2_7

7.1.1 Rack-Cylinder Welding Experiment and Measurement

The main dimensions of the examined cylindrical leg structure were 36,000 mm in length and 520 t in weight, as shown in Fig. 7.1, where the locations of measured sections are also depicted. The thickness of the cylinder was 110–115 mm and the material was grade of EH36. The dimensional features of the cross-section are shown in Fig. 7.2. Note that the outer and inner diameters were 4,200 mm and 3,980 mm, respectively. Figure 7.2 also shows the geometrical features of the welded joint. Four racks were located on the outer surface of the cylinder in symmetrical distribution with 17° away from the upright center line. As shown in Figs. 7.3 and 7.4, rack and cylinder were connected and fixed together with two 50-mm thickness plates. These two thick plates were previously welded to the rack with enough precision and strength. Then, they were welded to a cylinder by means of V-type welding groove and a total of 61 multi-pass welding seams were symmetrically and simultaneously welded. Shielded metal arc welding was employed. The welding conditions are summarized in Table 7.1. The preheating temperature from induction heating was controlled to be approximately 150 °C whereas the inter-pass temperature was controlled to be approximately 200 °C.

After the welding between rack and cylinder, the measurements of welding distortion showed that the cross-section shape of the cylinder presented a significant and evident change due to the welding shrinkage on the outer surface. In particular, the cross-section of the examined cylindrical leg structure was transformed from circle to elliptic as follows: the distance between the upper and lower edges in upright direction shrank approximately 20 mm, and the distance between the left and right edges in horizontal direction expanded approximately 7 mm. This type of geometrical change caused by welding influenced the interaction between the cylindrical leg structure and the working platform, and even the operation performance and efficiency for the whole jack-up rig. Figure 7.5 shows the cylinder straightness at four positions (right, left, down, and up points indicated in Fig. 7.2 at each measured section after welding; the negative deformation means inward shrinkage and positive deformation means outward expansion).

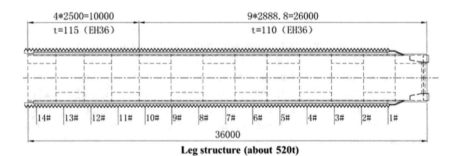

Fig. 7.1 Side view of leg structure and location of measured sections

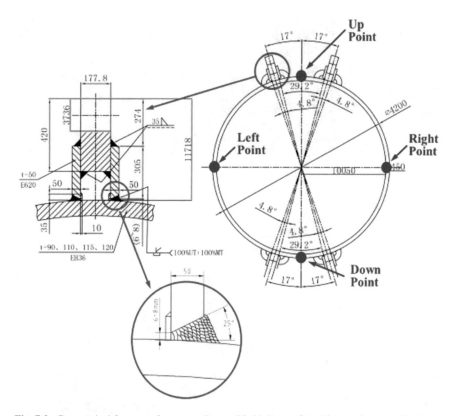

Fig. 7.2 Geometrical features of cross-section, welded joint configuration, and measured points

Fig. 7.3 Side view of welded joint design between rack and cylinder

The rigidity of the cylindrical leg structure was improved with the inner rib ring shown in Fig. 7.6a or the pillar stiffener shown in Fig. 7.6b to avoid welding distortion and ensure fabrication accuracy. Although this practical approach could prevent the welding distortion, as shown in Fig. 7.7 for the inner rib ring in actual fabrication, the manufacturing and procedure cost was increased. In addition, the total weight

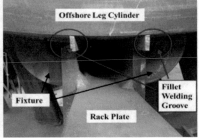

Fig. 7.4 Front view of welded joint design between rack and cylinder

Table 7.1 Welding condition for rack and cylinder welding

	Current (A)	Voltage (V)	Speed (mm/s)
Root/Cap Welding	190	23.5	3.17
Fill Welding	225	25	3.92

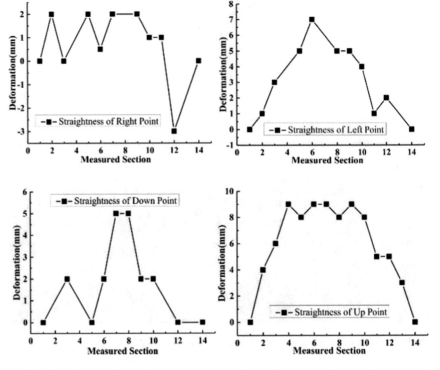

Fig. 7.5 Cylinder straightness of measured section after welding (minus: inward; plus: outward)

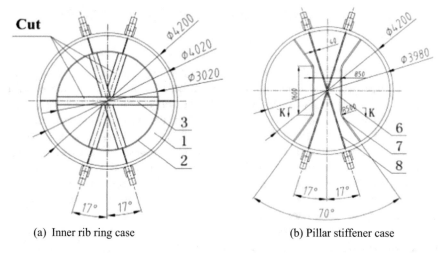

(a) Inner rib ring case   (b) Pillar stiffener case

Fig. 7.6 Design of rigidity reinforcement solution

Fig. 7.7 Actual fabrication with inner rib ring

was increased by approximately 40 t when some rigidity support components cannot be removed. We aimed at proposing an advanced technique not only to control the welding deformation but also to decrease the fabrication cost and weight of the cylindrical leg structure.

7.1.2 Welding Distortion Prediction with Efficient TEP FE Computation

For the multi-pass welded joint, large computer memory and long computing time were required during TEP FE analysis. An effective TEP FE computation with ISM and parallel computation was proposed and employed for welding deformation prediction. The solving domain of the FE model for the examined welded structure was divided into two regions during mechanical analysis. Both were updated by moving the heating source and by the procedure of filler metal added to the welding pass. The nodal temperature evaluated during thermal analysis was considered as the critical parameter to determine the size and location of a strong non-linear region.

To understand the mechanism of cross-section change behavior of the examined cylindrical leg structure after welding fabrication, a simplified cylinder model with bead-on-plate welding on the outer surface was previously investigated. Then, owing to the symmetrical feature and to reduce the computation consumption, a quarter FE model of the rack and cylinder welded structure was created to predict the welding deformation and compare it with experimental measurements.

7.1.2.1 Simplified Cylinder Model and Mechanism Clarification

Identical dimensions to those of the cylindrical leg structure without rack component were considered, as shown in Fig. 7.8. The total number of points and elements were 37,824 and 25,767, respectively. Multi-pass bead-on-plate welding lines were performed on the outer surface, and the length of the welding line was assumed to be 500 mm. Welding conditions are summarized in Table 7.1 and the material properties are shown in Fig. 7.9. All of them are identical to the experimental results.

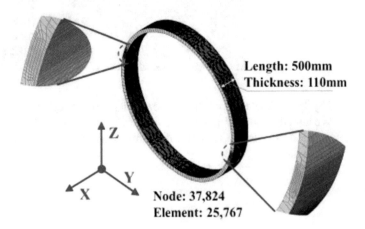

Fig. 7.8 Simplified cylinder model with bead-on-plate welding

Fig. 7.9
Temperature-dependent
material properties of EH36

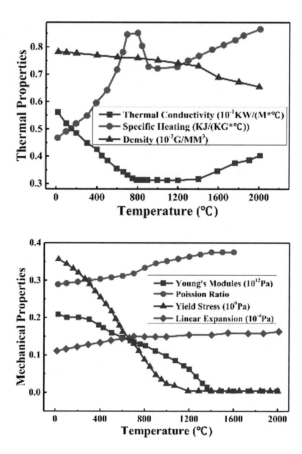

The inter-pass temperature is considered to be approximately 200 °C for subsequent FE computation.

Figure 7.10 shows the contour plotting of the highest temperature distribution and welded zone. Note that the welding arc presented a slightly shallower penetration in the thickness direction compared to the thickness of the cylinder. This type of temperature distribution caused by welding did not shrink the examined cylinder perimeter but influenced its shape. Figure 7.11 shows a comparison of the original cross-section shape of the examined cylinder, in yellow, with the deformed shape caused by welding, in orange. Note that the cross-section shape presented welding distortion with shrinkage in the upright direction and expansion in the horizontal direction.

From the above computational results and related explanation, the mechanism of welding distortion generated during rack and cylinder joining on the cross-section shape of the cylindrical leg structure can be clearly understood. To demonstrate the efficiency of the proposed methods during TEP FE computation, the computational time with the FE model shown in Fig. 7.8 are summarized and compared in Table 7.2.

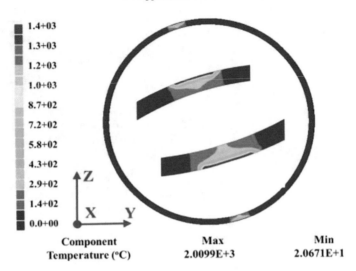

1.4+03	
1.3+03	
1.2+03	
1.0+03	
8.7+02	
7.2+02	
5.8+02	
4.3+02	
2.9+02	
1.4+02	
0.0+00	

Component	**Max**	**Min**
Temperature (°C)	**2.0099E+3**	**2.0671E+1**

Fig. 7.10 Contour plotting of the highest temperature distribution

Fig. 7.11 Comparison of cross-section shapes after welding (deformed scale: 50)

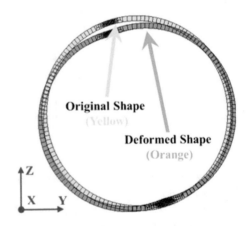

Original Shape
(Yellow)

Deformed Shape
(Orange)

Table 7.2 Comparison of computational efficiency with the proposed methods (Compiler: Intel Parallel Studio XE 2011, Platform: Ubuntu 14.04 OS and Dell PowerEdge T420)

	Time of thermal (s)	Time of mechanical analysis	
		Without ISM	With ISM
Conventional approach	920	108 h 16 min 54 s	32 h 6 min 47 s
With Intel MKL	494	21 h 28 min 39 s	20 h 54 min 18 s
With OpenMp parallel computation (10 cores)	424	108 h 52 min 47 s	32 h 41 min 54 s
With Intel MKL and OpenMp Parallel (10 cores)	240	20 h 24 min 2 s	19 h 33 min 30 s

In-house Fortran code was compiled with Intel Parallel Studio XE 2011 based on the platform of Ubuntu 14.04 OS and Dell PowerEdge T420 Server.

It can be concluded that the computing time during the thermal analysis can be significantly reduced to half using Intel Math Kernel Library (Intel MKL) or OpenMP parallel computation with 10 threads, and to quarter using both Intel MKL and OpenMP parallel computation with 10 threads. Taking into account the notable increase of TEP FE computation, the computational efficiency was compared not only for thermal analysis but also for mechanical analysis, as summarized in Table 7.2. Note that parallel computation with OpenMP can significantly reduce the computational time, which for mechanical analysis was reduced by 70% through a previous iterative substructure method and by 82% through a proposed iterative substructure method. This proposed method with parallel computation can reduce approximately 40% the computational time compared with the previous iterative substructure method.

7.1.2.2 Welding Distortion Prediction and Validation

To predict the welding deformation accurately and validate it with measured data, a quarter FE model of the rack and cylinder welded structures shown in Fig. 7.12 was used for the sake of geometrical symmetry and reduction of computational resources. Figure 7.12 shows 61 multi-pass welding seams with different colors belonging to 12 layers in V-type welding groove. The welding passes located at the left and right regions of the rack were welded symmetrically and simultaneously. The total number of points and elements were 122,859 and 114,350, respectively.

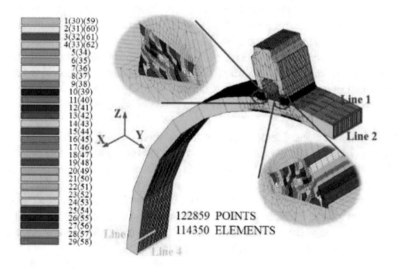

Fig. 7.12 Quarter FE model of rack and cylinder welded structures

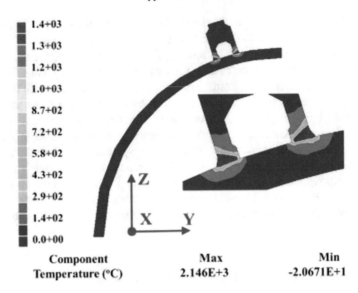

Component	Max	Min
Temperature (°C)	2.146E+3	-2.0671E+1

Fig. 7.13 Contour plotting of the highest temperature during the rack and cylinder welding (welded zone)

The horizontal direction of points on lines 1 and 2 and upright direction of points on lines 3 and 4 were all fixed to represent the symmetrical boundary condition. Rigid body motion was also constrained as mechanical boundary condition.

The welding condition is summarized in Table 7.1 whereas the temperature-dependent material properties of EH36 are shown in Fig. 7.9. Thermal analysis for multi-pass welding was carried out according to actual welding procedures and sequences during the experiment. Figure 7.13 shows the computed contour plotting of the highest temperature during the rack and cylinder welding. Note that the predicted welded zone was almost identical to the designed weld groove, and the arc welding heat source slightly penetrated the cylinder thickness. As mentioned above, this type of temperature distribution generated welding plastic strain in the outer surface and then influenced the cross-section shape of the examined cylindrical leg structure.

Appling the computed temperature profile as thermal load, mechanical analysis during welding was also carried out to compute the plastic strains and predict the welding deformation and residual stress. With a focus on the cross-section shape of the cylinder, Fig. 7.14a shows that the left edge presented a negative welding deformation in the horizontal direction, which means that the left edge expanded outward with respect to the origin. In addition, Fig. 7.14b shows that the upper edge also presented a negative welding deformation in the upright direction, which means that the upper edge shrank inward with respect to the origin.

Concerning the symmetrical characteristic, Fig. 7.15 shows the cross-section shape of the examined cylindrical leg structure after rack and cylinder welding.

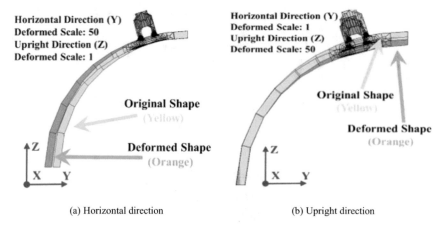

(a) Horizontal direction   (b) Upright direction

Fig. 7.14 Comparison of cross-section shape after rack and cylinder welding

Fig. 7.15 Predicted cross-section shape after rack and cylinder welding (deformed scale: 50)

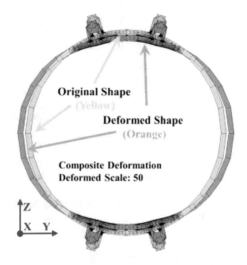

The original and deformed shapes are marked with yellow and orange colors. Note that the same deformation trends as in the experiments were obtained.

Although the computed welding deformation exhibited an identical tendency compared with experimental observation, it is still necessary to compare the magnitude of computed welding deformation with the corresponding measured data. Computed welding deformation of points on line 2 are indicated in Fig. 7.12. Note the shrinkage feature compared with the measurement with inward trend shown in Fig. 7.16a; good agreement was observed between both of them. Similarly, the computed welding deformation of points on line 4 with expansion feature shown in Fig. 7.12 was compared with the measurement with outward trend shown in Fig. 7.16b; good agreement was also observed. According to the comparison between

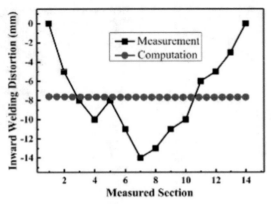

(a) Welding deformation of points on line 2 (inner part of upper edge)

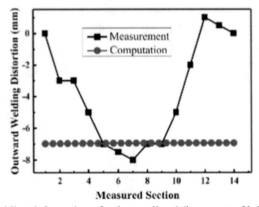

(b) Welding deformation of points on line 4 (inner part of left edge)

Fig. 7.16 Comparison of computed welding deformation and measured data (minus: inward; plus: outward)

measurements and computations, the measured welding distortions were located at the sections shown in Fig. 7.16 with a total length of 36,000 mm. The welding length of the quarter FE model of the rack and cylinder welded structures was 500 mm. The predicted welding distortion was consistent and could be used to approximate the welding distortion configuration of the weld structure FE model with actual length. Ignoring the influence of the welding end on the measurements, the middle region of the measured sections exhibited a uniform magnitude, which basically agrees with the computed results.

7.1.3 Welding Distortion Mitigation with Bead-on-Plate Techniques

According to the results presented above, TEP FE computation can predict not only the deformed shape trend but also the magnitude of welding deformation. The cross-section shape presented a significant and evident change owing to the arc welding on the outer surface and the generated plastic strain.

Although flame heating is usually employed for welding distortion mitigation with reverse deformation [1, 2], a variation of cross-sectional shape of the considered cylindrical leg structure after welding was dominantly caused by in-plane shrinkage rather than angular distortion. Thus, bead-on-plate welding on the inner surface was proposed to obtain much more transverse in-plane plastic strain rapidly, while plastic strain of the considered point was determined by the maximal temperature reached during the welding process and transverse plastic strains generated by the adjacent welding line were superposed [3]. When the bead-on-plate welding is performed on the inner surface with equivalent influence on the cross-section of the cylindrical leg structure, the fabrication accuracy influenced by welding may be controlled.

A new quarter FE model of the cylindrical leg structure with bead-on-plate welding on the inner surface is shown in Fig. 7.17. The welding conditions are summarized in Table 7.1, whereas the temperature-dependent material properties of EH36 are shown in Fig. 7.9. The welding procedures and symmetrical boundary condition are all identical to the TEP FE computation above described. Thermal and mechanical analyses were conducted with consideration of the bead-on-plate welding on the inner surface of the cylinder. Figure 7.18 shows the contour plotting of the highest temperature during the actual welding and bead-on-plate welding. Note that the bead-on-plate

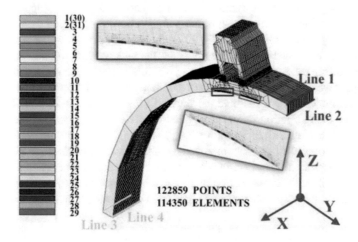

Fig. 7.17 Quarter FE model of rack and cylinder welded structure with bead-on plate welding on inner surface

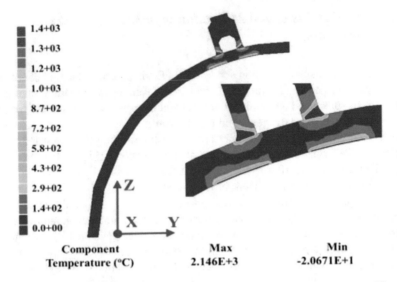

Fig. 7.18 Contour plotting of the highest temperature during the rack and cylinder welding and bead-on-plate welding

welding on the inner surface created a shallow welded zone and temperature distribution near the inner surface. This could not change the perimeter of the examined cylinder but reduced the influence of welding deformation during rack and cylinder welding on the cross-section shape of the cylindrical leg structure.

Furthermore, 30 bead-on-plate welding passes were applied on the inner surface and directly under the rack welding region. Figure 7.19a shows that welding deformation in the upright direction was reduced from 2.03 mm to 0.55 mm by approximately 72.9% and welding deformation in horizontal direction was reduced from 1.84 mm to 0.49 mm by approximately 73.4%. When 60 bead-on plate welding passes were applied on the inner surface and directly under the rack welding region, the welding distortion during rack and cylinder welding was totally mitigated, but additional and unnecessary welding deformation with 0.79 mm in horizontal direction and 0.95 mm in upright direction was generated, as shown in Fig. 7.19b. In addition, Table 7.3 summarizes the mitigation influence of bead-on-plate welding on the inner surface of the cylindrical leg with different welding passes.

It may be concluded that the employed technique with bead-on-plate welding on inner surface can significantly reduce the welding distortion during rack and cylinder welding and ensure the cross-section shape and fabrication accuracy of the cylindrical leg structure without complex processing schedule, and weight and cost increase.

Fig. 7.19 Trends of welding deformations with actual welding and bead-on-plate welding on inner surface

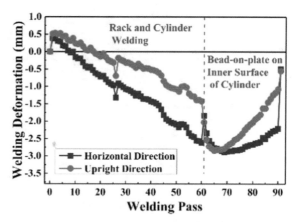

(a) Bead-on-plate with 30 welding passes

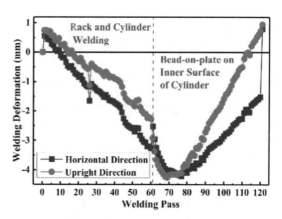

(b) Bead-on-plate with 60 welding passes

Table 7.3 Comparison of mitigation influence of bead-on-plate welding (minus: inward; plus: outward)

	Current	30 Bead-on-plate	60 Bead-on-plate
Upright direction	−2.03 mm	−0.55 mm (27%)	0.95 mm
Horizontal direction	−1.84 mm	−0.49 mm (26.6%)	0.79 mm

7.2   Welding Distortion Prediction and Mitigation Practice of Cantilever Beam Structure

The examined cantilever beam component, i.e., the so-called I section welded structure, which is the major structure of the jack-up jig, plays an essential role during

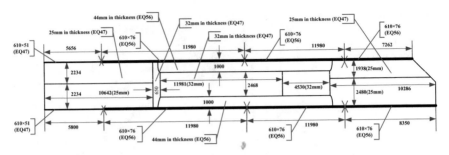

Fig. 7.20 Side view of the design drawing for examined cantilever beam components

the movement of cantilever beams. Figure 7.20 shows the side view of the design drawing of an I section welded structure in an examined cantilever beam, which typically consists of a top plate, web part, and bottom plate. The total length of the top plate was 36,878 mm and its width was 610 mm uniformly. This top plate was produced by four pieces of high tensile strength steel with respective dimensions of 5,656 mm × 610 mm × 51 mm (EQ 47), 11,980 mm × 610 mm × 76 mm (EQ 56), 11,980 mm × 610 mm × 76 mm (EQ 56), and 7,262 mm × 610 mm × 76 mm (EQ 56), which were joined with butt welding. Almost the same configuration was applied for the fabrication of the bottom plate, as shown in Fig. 7.20, with the length changed to be 38,110 mm and the lengths of the four pieces changed to be 5,800 mm, 11,980 mm, 11,980 mm, and 8,350 mm, respectively. Concerning the web part, it was composed of several pieces of steel plates with different sizes, thicknesses, and materials, as shown in Fig. 7.20. The web part was also produced by butt welding of steel plates with the same or different thicknesses.

The materials EQ 47 and EQ 56 were produced according to the American Bureau of Shipping standard. Both materials exhibit super high strength quenched and tempered steel grades. They are usually used in offshore drilling platforms, shipping marine engineering, and hull structural projects. The chemical compositions of EQ 47 and EQ 56 are summarized in Table 7.4, and their mechanical properties are shown in Table 7.5.

Three typical welded joints in the I section welded structure are (a) butt welded joint with the same thickness, (b) butt welded joint with different thicknesses, and (c) fillet welded joint; they are depicted in Fig. 7.21. To join the thick plates, SAW was employed for downward welding with a diameter of welding wire of 4.0 mm. FCAW with CO_2 shielding gas was applied for horizontal position welding with a diameter of welding wire of 1.2 mm. The preheating temperature was achieved at

Table 7.4 Chemical composition of EQ 47 and EQ 56

	C% Max	Si% Max	Mn% Max	P% Max	S% Max
Equation 47	0.20	0.55	1.70	0.030	0.030
Equation 56	0.20	0.55	1.70	0.030	0.030

Table 7.5 Mechanical properties of EQ 47 and EQ 56

	Yield strength (MPa)	Tensile strength (MPa)	Elongation (%)	Impact energy (J at − 40 °C)
Equation 47	460	570-720	17	46
Equation 56	550	670-835	16	55

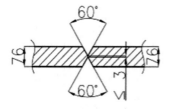

(a) Butt welded joint with the same thickness

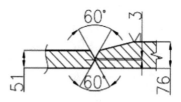

(b) Butt welded joint with different thicknesses

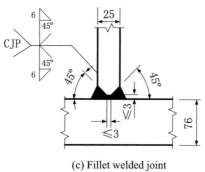

(c) Fillet welded joint

Fig. 7.21 Design drawings of typical welded joints

150 °C when the plate was thicker than 38 mm; otherwise, it was achieved at 200 °C. The temperature between welding passes was controlled to be 150 °C when the plate was thinner than 38 mm; otherwise, it was 230 °C. Other welding conditions are summarized in Table 7.6 for butt welding and Table 7.7 for fillet welding.

Following the conventional assembly processes, the I section welded structure in the examined cantilever beam can be fabricated as follows:

(1) The bottom plate is produced by sequentially joining the four pieces of steel plates on the platform.
(2) The web part is slightly more complex. It is generally divided into three segments: front, middle, and end sections. These three sections are produced separately, and then they are assembled together through butt welding.
(3) Subsequently, the web part is lifted to become vertically positioned onto the bottom plate, and fillet welding is employed to join the web part and bottom plate.

Table 7.6 Welding condition for butt welding

	Thickness (mm)	No. of welding pass	Current (A)	Voltage (V)	Velocity (mm/s)
B1	25 (left) + 25 (right)	5 (1G) + 8 (4G)	220–260(1G) + 185–225(4G)	22–26(1G) + 21–25(4G)	150–270(1G) + 125–205(4G)
B2	32 (left) + 32 (right)	6 (1G) + 8 (4G)	220–260(1G) + 185–225(4G)	22–26(1G) + 21–25(4G)	150–270(1G) + 125–205(4G)
B3	76 (left) + 76 (right)	25 (1G) + 35 (4G)	220–260(1G) + 185–225(4G)	22–26(1G) + 21–25(4G)	150–270(1G) + 125–205(4G)
B4	25 (left) + 32 (right)	5 (1G) + 8 (4G)	220–260(1G) + 185–225(4G)	22–26(1G) + 21–25(4G)	150–270(1G) + 125–205(4G)
B5	25 (left) + 44 (right)	5 (1G) + 8 (4G)	220–260(1G) + 185–225(4G)	22–26(1G) + 21–25(4G)	150–270(1G) + 125–205(4G)
B6	32 (left) + 44 (right)	6 (1G) + 8 (4G)	220–260(1G) + 185–225(4G)	22–26(1G) + 21–25(4G)	150–270(1G) + 125–205(4G)
B7	51 (left) + 76 (right)	15 (1G) + 20 (4G)	220–260(1G) + 185–225(4G)	22–26(1G) + 21–25(4G)	150–270(1G) + 125–205(4G)

Table 7.7 Welding condition for fillet welding

	Thickness (mm)	No. of welding bead	Current (A)	Voltage (V)	Velocity (mm/s)
T1	51 (plate) + 25 (web)	11 (2G)	220–260	22–26	150–270
T2	76 (plate) + 25 (web)	11 (2G)	220–260	22–26	150–270
T3	76 (plate) + 32 (web)	14 (2G)	220–260	22–26	150–270
T4	76 (plate) + 44 (web)	18 (2G)	220–260	22–26	150–270

1G: downward welding; 2G: horizontal position welding; 4G: overhead position welding

(4) Almost the same welding procedure used for bottom plate joining is carried out to produce the top plate.

(5) Finally, the top plate is placed on the platform, and the welded structure with bottom plate and web part are placed onto the top plate. Fillet welding is employed again for joining the top plate to the already existing T welded structure.

Owing to the double-side welding for welding distortion control, the welded plate may be joined with overhead position welding or be lifted and reversed during the assembly process. Moreover, the fillet welding is carried out with horizontal position welding. Alternate welding pass is sequentially achieved at the left and right sides.

7.2.1 Establishment of Inherent Deformation Database

To use elastic FE analysis with inherent deformation to predict welding distortion of welded structures, inherent deformation must be previously evaluated. There are two ways to determine the values of inherent deformation. One is based on experimental measurement, and the other is implemented with thermal elastic-plastic FE computation. The later was employed in this study. As discussed in the previous section, SAW with welding condition shown in Table 7.6 was employed for butt welded joint, whereas FCAW using CO_2 shielding gas with welding condition shown in Table 7.7 was employed for fillet welded joint. Different welding methods and procedures were employed for different types of welded joints. However, from the perspective of physical phenomenon, a welding arc in different welding procedures produces a moving source of local high intensity power in the plate and generates a sharp thermal profile in the weld pool, HAZ, and base metal. Therefore, a body heat source can be employed to model SAW and FCAW with different geometrical shapes during thermal elastic-plastic FE analysis. A birth-death element technique was employed during thermal elastic-plastic FE analysis of multi-pass welding to activate the dummy elements when the corresponding welding pass was applied.

7.2.1.1 Solid-Element Models of Typical Welded Joints

According to Fig. 7.21, and Tables 7.6 and 7.7, a series of solid-element models were created. These models were used to conduct thermal elastic-plastic FE analysis later. A welded joint model with 400 mm in width was extracted from the welded structure of examined cantilever beam, and a free case for the assembling of these typical welded joints were then examined. Rigid body motion was employed as boundary condition for inherent deformation evaluation.

Figure 7.22 shows the FE models with solid elements of butt welded joints with the same thickness. Figures 7.22a and b depict the welded joints in the web part whereas Fig. 7.22c shows the top/bottom plate. The distribution and geometrical information of each welding pass were enlarged in Fig. 7.22, and the individual welding passes were marked with different colors. The total number of nodes and elements in each FE model of butt welded joints with same thickness are also indicated in Fig. 7.22.

Similarly, FE models with solid elements of butt welded joints with different thicknesses and fillet welded joints are shown in Figs. 7.23 and 7.24, respectively. The distribution and geometrical information of each welding pass are also enlarged in these figures, and individual welding passes are marked with different colors. The total number of nodes and elements in each FE model of butt welded joints with different thicknesses are indicated in Fig. 7.23, and those in each FE model of fillet welded joints are indicated in Fig. 7.24.

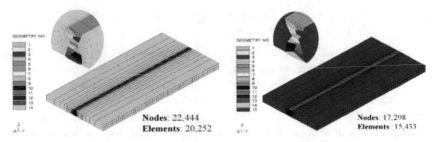

(a) FE model of butt welded joint B1 (25 mm + 25 mm) (b) FE model of butt welded joint B2 (32 mm + 32 mm)

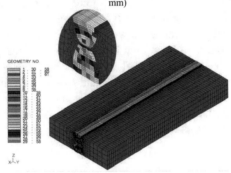

(c) FE model of butt welded joint B3 (76 mm + 76 mm)

Fig. 7.22 Solid-element models of butt welded joint with the same thickness

7.2.1.2   Transient Temperature Results

During thermal computation with thermal elastic-plastic FE analysis, the initial temperature is often assumed to be room temperature. A body heat source with uniform power density Q (w/m³: welding arc energy/volume of body heat source) was employed to model the heat source of welding arc in this study. Moreover, heat losses due to convection and radiation were also taken into account in the finite element model. Combined convection and radiation boundary conditions generated a boundary flux q (w/m²) on all the external surfaces [4]. The time step during heating was defined according to the total computing time, computed accuracy, and convergence. After the heating was completed, the time-step length was exponentially increased until the sum of computing time reached a defined maximum examined time at which the cooling process was rapidly considered for improving the computed efficiency. Three typical welded joints, namely (a) butt welding with same thickness, (b) butt welding with different thicknesses, and (c) fillet welding were taken as examples to demonstrate the transient thermal analysis. Figure 7.25 shows the transient temperature distribution resulting from thermal analysis when the welding arc was passing 1/3 of the position of the welding line.

Figure 7.25(a) shows the transient temperature distribution in the butt welded joint assembled from plates with a same thickness of 25 mm. When two plates with

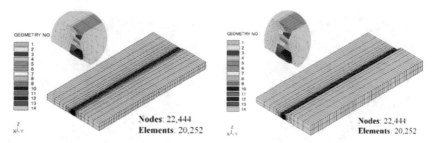

(a) FE model of butt welded joint B4 (25 mm + 32 mm) (b) FE model of butt welded joint B5 (25 mm + 44 mm)

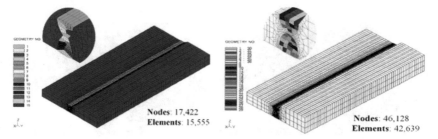

(c) FE model of butt welded joint B6 (32 mm + 44 mm) (d) FE model of butt welded joint B7 (51 mm + 76 mm)

Fig. 7.23 Solid-element models of butt welded joint with different thicknesses

32 mm and 44 mm in thickness were welded to produce a butt welded joint, the transient temperature distribution was obtained; it is shown in Fig. 7.25b. Consider a fillet welded joint as the one shown in Fig. 7.25c as an example. It was produced by vertically welding a web with 25 mm in thickness onto a plate with 76 mm in thickness. Figure 7.25c also shows the transient temperature distribution when the welding arc was passing 1/3 of the position of the welding line at the right side.

7.2.1.3 Plastic (Inherent) Strain Distribution on Cross-Section

For mechanical computation, the initial stress is often assumed to be stress free, and the elastic-plastic problem is solved using a time marching scheme with time-step lengths extracted from thermal computation. In terms of boundary conditions, the welded joint is free to deform but rigid body motion is established as fixed as a constraint. The welded joints discussed in Fig. 7.25 were examined again to investigate the longitudinal and transverse plastic strains caused by welding. Owing to the extremely different self-constraints on welding bead in longitudinal and transverse directions, the features of distribution and magnitude of these plastic strains on one cross-section were studied from the computed results. In particular, the region of longitudinal plastic strain was wider than that in transverse direction. However,

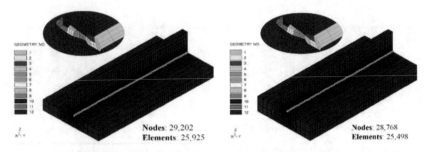

(a) FE model of fillet welded joint T1 (51 mm + 25 mm) (b) FE model of fillet welded joint T2 (76 mm + 25 mm)

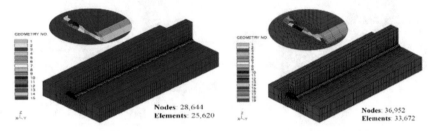

(c) FE model of fillet welded joint T3 (76 mm + 32 mm) (d) FE model of fillet welded joint T4 (76 mm + 44 mm)

Fig. 7.24   Solid-element models of fillet welded joint

the magnitude of longitudinal plastic strain was much smaller compared to that in transverse direction [5].

Figures 7.26 and 7.27 show the distributions of longitudinal and transverse plastic strains on the central cross-section after cooling down for butt welded joint B1(two plates with same thickness of 25 mm) and B6 (two plates with thicknesses of 32 and 44 mm), respectively. Note that the aforementioned features can be clearly distinguished. For the fillet welded joint shown in Fig. 7.28, the same features of plastic strains were obtained. The web plate with 25 mm in thickness was vertically welded onto the flange plate with 76 mm in thickness. Complete joint penetration was carried out to ensure suitable welding and mechanical performance in fillet welding. Therefore, the distributions of longitudinal and transverse plastic strains appeared near the welding passes and just throughout the web plate.

7.2.1.4   Inherent Deformation Database

Given that the plastic strain is the dominant component of inherent strain after cooling down, the values of inherent deformation can be evaluated in theory by substituting them into the definitions given by Eq. (2.8). However, the thickness in Eq. (2.8) denotes the thickness of welded joint [6], and the aforementioned method can be

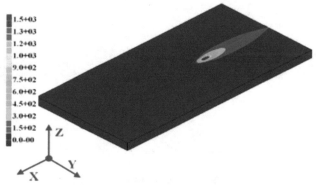

(a) Transient temperature distribution of butt welded joint (B1: 25 mm + 25 mm)

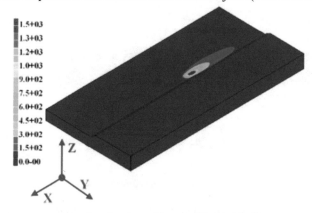

(b) Transient temperature distribution of butt welded joint (B6: 32 mm + 44 mm)

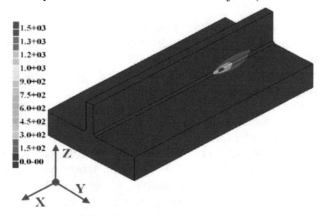

(c) Transient temperature distribution of fillet welded joint (T2: 76 mm + 25 mm)

Fig. 7.25 Transient temperature distribution of thermal analysis

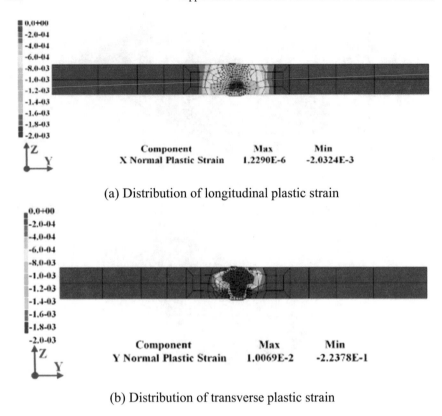

(a) Distribution of longitudinal plastic strain

(b) Distribution of transverse plastic strain

Fig. 7.26 Distribution of plastic strains of butt welded joint (B1: 25 mm + 25 mm)

used for the butt welded joints with full penetration, such as the welded joints with same thickness shown in Fig. 7.22 and with different thicknesses shown in Fig. 7.23. The evaluated values of inherent deformation for all examined butt welded joints in this study are summarized in Table 7.8 for the same thickness plates and Table 7.9 for different thicknesses plates.

For the fillet welded joint, it is difficult to distinguish the thickness of the welded joint. Inherent deformations can be evaluated according to their individual physical behavior. Longitudinal inherent shrinkage can be converted to tendon force for eliminating the influence of thickness of the welded joint, which is a result of strong self-constraint in longitudinal direction [6, 7]. Owing to the physical behavior such that weak self-constraint exists in the transverse direction, the transverse inherent shrinkage and bending moment was evaluated directly by welding displacement, as summarized in Table 7.10. In addition, for many typical welded joints such as butt and fillet welding, longitudinal inherent bending moment is usually ignored owing to its much smaller magnitude.

Furthermore, the inherent deformations listed in Tables 7.8, 7.9, and 7.10 result from integration of inherent strain (almost plastic strain), which is determined by

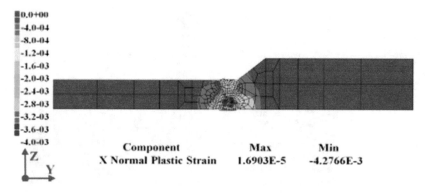

(a) Distribution of longitudinal plastic strain

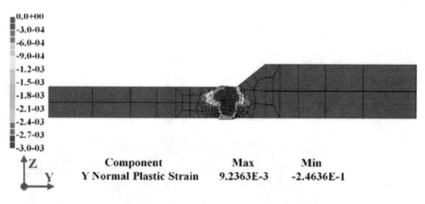

(b) Distribution of transverse plastic strain

Fig. 7.27 Distribution of plastic strains of butt welded joint (B6: 32 mm + 44 mm)

the welding condition, material properties, type of welded joint, etc. However, it is independent of the examined whole welded structure.

7.2.2 Welding Distortion Prediction with Elastic FE Analysis

Following the dimension and geometrical characteristics of the I section welded structure in the examined cantilever beam shown in Fig. 7.20, an FE model with shell elements was created, as shown in Fig. 7.29. This model comprised 18 parts, 3,000 nodes, and 2,446 shell elements. The adjacent parts with different colors were welded together according to the conventional assembly processes discussed in the previous section.

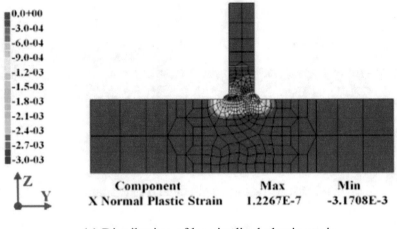

(a) Distribution of longitudinal plastic strain

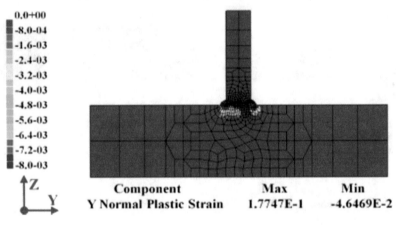

(b) Distribution of transverse plastic strain

Fig. 7.28 Distribution of plastic strains of fillet welded joint (T2: 76 mm + 25 mm)

Table 7.8 Evaluated inherent deformations of butt welded joint with same thickness

	Thickness (mm)	Longitudinal inherent shrinkage (mm)	Transverse inherent shrinkage (mm)	Transverse inherent bending (rad)
B1	25 (left) + 25 (right)	0.05647	0.94242	0.00071002
B2	32 (left) + 32 (right)	0.05512	0.82274	0.00044086
B3	76 (left) + 76 (right)	0.05894	0.98362	0.0004518

Table 7.9 Evaluated inherent deformations of butt welded joint with different thicknesses

	Thickness (mm)	Longitudinal inherent shrinkage (mm)	Transverse inherent shrinkage (mm)	Transverse inherent bending (rad)
B4	25 (left) + 32 (right)	0.04865	0.83532	0.00061510
B5	25 (left) + 44 (right)	0.04834	0.73519	0.00053906
B6	32 (left) + 44 (right)	0.04481	0.70238	0.00037117
B7	51 (left) + 76 (right)	0.07304	0.538592	0.00164520

Table 7.10 Evaluated inherent deformations of fillet welded joint

	Thickness (mm)	Longitudinal inherent shrinkage (mm)	Transverse inherent shrinkage (mm)	Transverse inherent bending (rad)
T1	51 (plate) + 25 (web)	0.03079	0.497186 (plate)	0.00348360(plate)
			1.471531 (web)	0.00058128 (web)
T2	76 (plate) + 25 (web)	0.01954	0.36315 (plate)	0.00112780(plate)
			1.426776 (web)	0.0006423 (web)
T3	76 (plate) + 32 (web)	0.0239	0.45544 (plate)	0.0013968(plate)
			1.19447 (web)	0.00036021 (web)
T4	76 (plate) + 44 (web)	0.0358	0.331398 (plate)	0.0020386(plate)
			0.851749 (web)	0.00031911 (web)

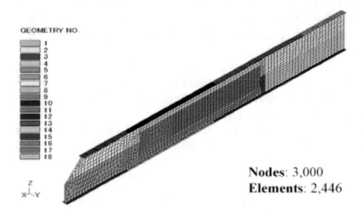

GEOMETRY NO.

Nodes: 3,000
Elements: 2,446

Fig. 7.29 Shell-element model of I section welded structure

Using the previously evaluated inherent deformations and considering the fitting process before full welding with interface element, elastic FE analysis was carried out for welding distortion prediction. These inherent deformations were applied into the welding lines between the two parts to be welded.

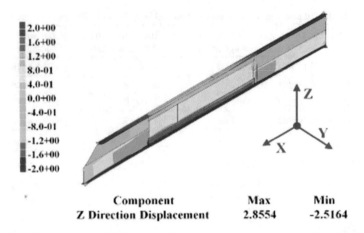

Component	Max	Min
Z Direction Displacement	**2.8554**	**-2.5164**

Fig. 7.30 Contour distribution of Z displacement of examined I section welded structure after assembly (deformed scale: 100)

The computed results shown in Fig. 7.30 revealed that the top and bottom plates deformed downward and upward, respectively. Note also that the maximum deflection occurred in the middle part of the I section welded structure. This type of computed distribution of out-of-plane welding distortion can be explained in the following terms:

(1)   During the welding of thick plates, symmetrical welding is usually employed to reduce the potential transverse bending or angular distortion. Therefore, less magnitude of out-of-plane welding distortion in both butt and fillet welded joints are generated.

(2)   Owing to the negligible transverse bending caused by welding, out-of-plane welding distortion of the examined I section welded structure was dominantly produced by in-plane shrinkage. Consequently, the top and bottom plates almost deformed uniformly.

(3)   In the middle part of the I section welded structure, there were more welding lines than those in the front and end parts. Larger in-plane shrinkages in longitudinal and vertical directions were produced during the assembly process of the web part.

To examine the flatness of the cantilever beam closely, the deflections of edges of top and bottom plates were measured. The sum of magnitudes along the longitudinal direction are shown in Fig. 7.31. It may be concluded that vertical in-plane shrinkage generated by the welding line in the web part causes sharp distribution of out-of-plane welding distortion, as shown in Fig. 7.31.

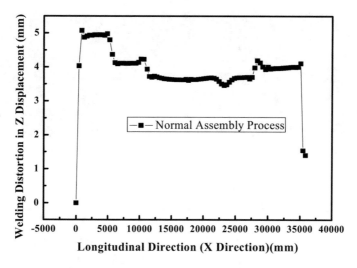

Fig. 7.31 Sum of magnitudes of deflection of edges at top and bottom plates

7.2.3 Mitigation Implementation with Practical Techniques

To reduce the welding distortion during the fabrication of the I section welded structure in the examined cantilever beam to an acceptable level, some mitigation techniques were practiced during the actual manufacturing. In this study, three different types of mitigation techniques were practically implemented and examined through computational analysis, namely (1) application of inverse distortion to the top and bottom plates before fillet welding, (2) assembling with improved welding sequence, and (3) external constraint with tack welding and mechanical fixture.

7.2.3.1 Application of Inverse Distortion

Inverse distortion is always employed to offset the transverse bending for mitigation of out-of-plane welding distortion in actual fabrication. Figure 7.32 shows the backing structure and pressure equipment for inverse distortion generation. In particular, the backing structure shown in Fig. 7.32a supports the center part for it to be higher than the edge parts. Then, the pressure equipment and constraint shown in Fig. 7.32b jointly bend the plate downward. By controlling the height of the center part of the backing structure, appropriate magnitude of inverse distortion opposed to transverse bending in the welded plates was generated. In this study, the magnitude of inverse distortion was approximately 10 mm due to the applied mechanical device.

After the generation of inverse distortion before welding, only in-plane inherent shrinkages were applied for welding distortion prediction through elastic FE analysis. The computed results depicted in Fig. 7.33 show a similar contour distribution of out-of-plane welding distortion to that shown in Fig. 7.30. However, less magnitude

(a) Backing structure   (b) Pressure equipment and constraint

Fig. 7.32 Generation of inverse distortion before welding

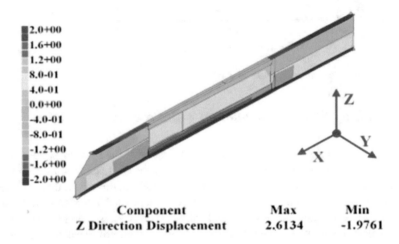

Component	Max	Min
Z Direction Displacement	2.6134	-1.9761

Fig. 7.33 Contour distribution of Z displacement of the examined I section welded structure considering inverse distortion (deformed scale: 100)

of out-of-plane welding distortion was obtained, as shown in Fig. 7.34; it was reduced by approximately 20% through application of inverse distortion.

Although the out-of-plane welding distortion can be reduced with application of inverse distortion, it can also be concluded that transverse bending was not the dominant reason in the generation of out-of-plane welding distortion at both top and bottom plates in this study, as discussed above. This welding distortion was mainly generated by the in-plane shrinkage of the web part.

7.2.3.2 Assembly with Improved Welding Sequence

According to the prescribed welding sequence, the web parts were welded together and then joined onto the top and bottom plates with fillet welding. There were two difficulties during the actual fabrication:

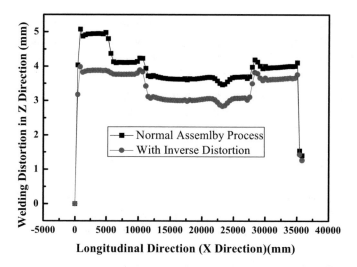

Fig. 7.34 Comparison of sum of magnitudes of deflection of edges of top and bottom plates considering inverse distortion

(1) A large magnitude of out-of-plane welding distortion of the web parts was generated, thereby further increasing the cost and straightening effort. Advanced fabrication consists of assembling the upper/lower web parts and then joining them with the top/bottom plates for enhancing their stiffness. Considering the welding distortion after assembling the web parts, the resulting out-of-plane welding distortion was predicted. The contour distribution is shown in Fig. 7.35. Note that the web part presented a large amount of bending distortion.

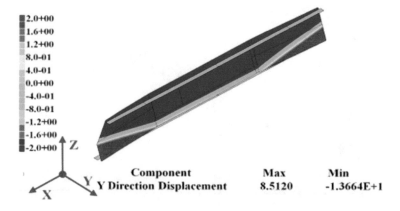

Fig. 7.35 Contour distribution of Y displacement of examined web part (deformed scale: 100)

(2)   For thick-plate welding, symmetrical welding was always employed as
       discussed above. It was easy to achieve when the upper and lower web parts
       were vertically welded. Concerning the large web part, it was difficult to move
       and lift in the workspace efficiently.

   Therefore, an improved welding sequence was applied in which the upper web
part was welded with the top plate and the lower web part was welded with the
bottom plate simultaneously; then, the upper and lower parts were joined through
butt welding, as shown in Fig. 7.36. This sequence was proposed to overcome the
aforementioned difficulties. The out-of-plane welding distortion of the web part was
subsequently examined to assess the improved welding sequence as follows:

(1)   The lower web part and bottom plate were assembled separately. Then, they
       were fitted and the lower web part was vertically welded onto the bottom plate
       with fillet welding. Figure 7.37 shows the overall welding distortion in lateral

(a) Finished lower web part   (b) Upper and lower parts before fitting and full welding

Fig. 7.36 Improved welding sequence to assemble the examined I section welded structure

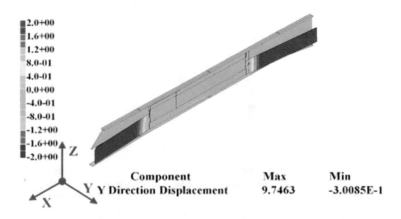

Fig. 7.37 Welding distortion during the fitting stage before full welding between the bottom plate
and lower web part (deformed scale: 100)

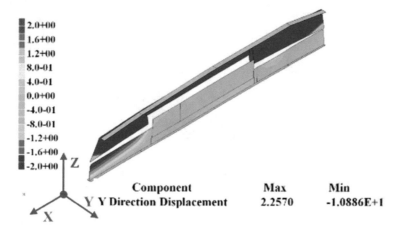

Component	Max	Min
Y Direction Displacement	2.2570	-1.0886E+1

Fig. 7.38 Welding distortion during the fitting stage before full welding between the top plate and upper web part (deformed scale: 100)

direction during the fitting stage before full welding between the bottom plate and lower web part.

(2) Simultaneously, the upper web part and top plate were assembled separately; then, they were fitted and the upper web part was vertically welded onto the top plate with fillet welding. Figure 7.38 shows the overall welding distortion in lateral direction during the fitting stage before full welding between the top plate and upper web part. Note that the bottom plate and lower web part were already welded.

(3) After finishing the upper and lower parts, they were fitted and joined through butt welding. Figure 7.39 shows the overall welding distortion in lateral direction during the fitting stage before full welding between the upper and lower parts.

The welding distortions (Y direction in the global coordinate system) of top and bottom welding lines of fillet welding were measured and compared, as shown in Fig. 7.40. Note that the welding distortion with the improved welding sequence was significantly eliminated, thereby reducing the time and cost for correction and straightening during the fitting stage. However, assembly with the improved welding sequence had less influence on the final magnitude of out-of-plane welding distortion of top and bottom plates, as shown in Fig. 7.41. Flatness had a strong effect on the movement performance of the examined cantilever beam.

The applied inherent deformations and fitting procedure were identical for the different welding sequences. However, the stiffness of the already assembled welded structure was totally different. It influenced the final dimensional accuracy caused by remaining (deformations generated by previous welding procedures were partially corrected during the fitting procedure, but a fraction of them remained) and current welding distortion.

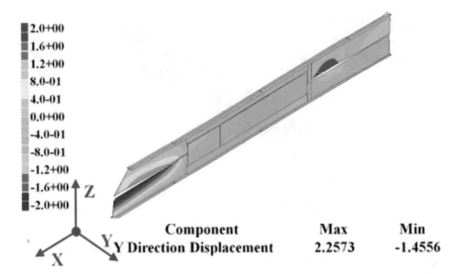

Component	Max	Min
Y Direction Displacement	2.2573	-1.4556

Fig. 7.39 Welding distortion during the fitting stage before full welding between the top and bottom parts

7.3   Conclusions

Two welded structures for jack-up drilling rig fabrication were examined. Welding distortion prediction and mitigation with advanced computational approaches were demonstrated. Elastic FE analysis with inherent deformation and interface element constitutes an effective and practical computational approach for welding distortion investigation in the fabrication of large-scale welded offshore structures. The following conclusions can be drawn:

(1)   The variation behavior of cross-section shape for the considered cylindrical leg structure during rack and cylinder welding was experimentally observed. Moreover, the generation mechanism of cross-section shape variation after welding was clarified by computational investigation.

(2)   Fabrication accuracy was investigated with a quarter FE model of actual rack and cylinder welded structures through proposed efficient TEP FE computation. Good agreement was achieved between the computed welding deformation and the corresponding measurements.

(3)   Bead-on-plate welding with multi-pass on the inner surface of the cylinder was numerically examined. The welding deformations generated by rack and cylinder welding in both horizontal and upright directions can be significantly reduced to ensure the fabrication accuracy without weight and cost increase.

(4)   For the welded structure assembled from thick plates, transverse bending or angular distortion of each welded joint has a less contribution to the final dimensional accuracy of the whole welded structure owing to the symmetrical welding and large self-stiffness. In-plane shrinkage near the welding line is the

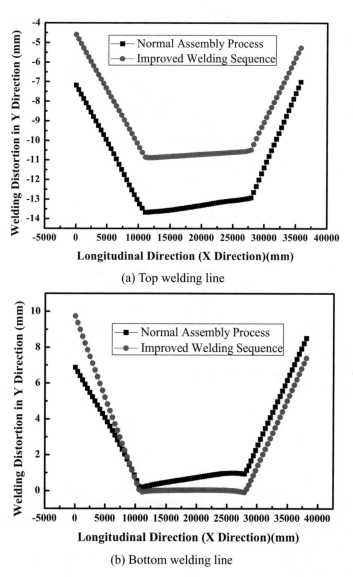

(a) Top welding line

(b) Bottom welding line

Fig. 7.40 Welding distortions in Y direction of top and bottom welding lines during the fitting stage before fillet welding

dominant reason in the generation of the final dimensional accuracy achieved in the fabrication of welded structures with thick plates.

(5) Transverse bending or angular distortion of each welded joint can be efficiently corrected by application of inverse distortion. Their magnitude can be evaluated by numerical analysis before the actual fabrication, but the applied inverse distortion is determined by the mechanical tool employed.

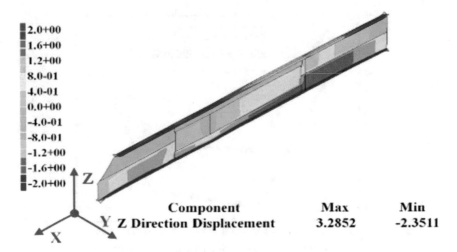

Fig. 7.41 Contour distribution of Z displacement of examined I section welded structure with improved welding sequence (deformed scale: 100)

(6)   The proposed welding sequence not only strongly influences the dimensional accuracy during the fitting and welding procedures owing to the stiffness of already assembled welded structures, but also influences the production schedule owing to the lifting feasibility of welded assemblies in actual manufacturing.

References

1. Wang J, Rashed S, Murakawa H, Luo Y (2013) Numerical prediction and mitigation of out-of-plane welding distortion in ship panel structure by elastic FE analysis. Mar Struct 34:135–155
2. Ma N, Wang J, Okumoto Y (2016) Out-of-plane welding distortion prediction and mitigation in stiffened welded structures. Int J Adv Manuf Technol 84(5–8):1371–1389
3. Luo Y, Murakawa H, Ueda Y (1997) Prediction of welding deformation and residual stress by elastic FEM based on inherent strain (report I): mechanism of inherent strain production. Trans JWRI 26(2):49–57
4. Wang JH, Luo H (2000) Prediction of welding deformation by FEM based on inherent strains. J Shanghai Jiaotong Univ (Engl. Editor. Board) 5 (2):83–87
5. Yang YP, Castner H, Kapustka N (2011) Developmeng of distortion modeling methods for large welded structures. J Ship Prod Des 27(1):26–34
6. Zhang L, Michaleris P, Marugabandhu P (2007) Evaluation of applied plastic strain methods for welding distortion prediction. J Manuf Sci Eng 129(6):1000–1010
7. Wang JC, Yuan H, Ma N, Murakawa H (2016) Recent research on welding distortion prediction in thin plate fabrication by means of elastic FE computation. Mar Struct 47:42–59

Printed in the United States
by Baker & Taylor Publisher Services